Sasquatch/Bigfoot
and the
Mystery of the Wild Man

Jean-Paul Debenat, PhD

Sasquatch/Bigfoot and the Mystery of the Wild Man

Cryptozoology & Mythology in the Pacific Northwest

**Translated by Paul LeBlond, PhD
Edited by Christopher L. Murphy**

ISBN 978-0-88839-685-3 [2010 print]
ISBN 978-0-88839-185-8 [2018 reprint]
ISBN 978-0-88839-686-0 [e-book]
Copyright © 2009 Jean-Paul Debenat

Cataloging in Publication Data

Debenat, Jean-Paul
 Sasquatch/Bigfoot and the mystery of the Wild Man :
 cryptozoology & mythology in the Pacific Northwest / Jean-Paul Debenat ;
 translated by Paul LeBlond ; edited by Christopher L. Murphy.

Translation of: Sasquatch et le mystère des hommes sauvages.
Includes bibliographical references and index.
Issued also in electronic format.
ISBN 978-0-88839-685-3

1. Sasquatch. 2. Wild men. 3. Indian mythology—Northwest, Pacific.
4. Cryptozoology—Northwest, Pacific. I. LeBlond, P.H. II. Murphy,
Christopher L. (Christopher Leo), 1941– III. Title.

QL89.2.S2D4313 2009   001.944   C2009-902395-4

All rights reserved. No part of this publication may be reproduced, stored in a retrieval system or transmitted, in any form or by any means, electronic, mechanical, photocopying, recording, or otherwise, without the prior written permission of Hancock House Publishers.

Printed in the USA

FRONT COVER
A reconstruction of bigfoot, completed in 2000 by sculptor Emmanuel Janssens-Casteels for Prof. Franco Tassi, director of the National Park of the Abruzzes, Italy, for display in a museum dedicated to mysterious animals and cryptozoology.

Crypto Editions is an imprint of Hancock House Publishers

Published simultaneously in Canada and the United States by

HANCOCK HOUSE PUBLISHERS LTD.
19313 Zero Avenue, Surrey, B.C. Canada V3Z 9R9
(604) 538-1114 Fax (604) 538-2262

HANCOCK HOUSE PUBLISHERS
#104-4550 Birch Bay Lynden Rd, Blaine, WA U.S.A. 98230-5005
(800) 938-1114 Fax (800) 983-2262

Website: **www.hancockhouse.com**
Email: **sales@hancockhouse.com**

Dedication

This book is dedicated to my wife Marie-Agnès whose patience and expert secretarial skills were absolutely indispensable.

Acknowledgements

My heartfelt thanks to the many people who contributed to this work, in particular:

Robert Alley (USA), Thelma Ashamire (USA), Richard Barbre (USA), Dmitri Bayanov (Russie), Joe Beelart (USA), Ralph Bennett (USA), Michel Besnier (France), Miggs Bodie (Australie), Fred Bradshaw (USA), John A. Brown (USA), Rob Butler (USA), Peter Byrne (USA), Ray Crowe (USA), John DeGraaf (USA), Nick Di Martino (USA), Dayton Edmonds (USA), Father Robert (Bob) Erickson (USA), Adeline Fredin (USA), Elizabeth Freeman (USA), Ed Fusch (USA), Chantal Gérard-Landry (France), Dr. Patrick Germanèse (France), Lauric Guillaud (France), John Green (Canada), John Heath-Stubbs (Grande-Bretagne), Dr. Bernard Heuvelmans (France), Vi Hilbert (USA), Bill Holm (USA), R. C. Hoover (USA), Jérôme Houdin (France), Carol Howell (USA), Emmanuel Janssens-Casteels (Belgique), Eric Joye (Belgique), Ken et Jo Jackson (Grande-Bretagne), Andy Joseph (USA), Dr. Marie-Jeanne Koffmann (France/Russie), Grover Krantz (USA), Paul LeBlond (Canada), André Lemelin (Canada), Christian Le Noël (France), Alika Lindbergh (France), Carol McMillan (USA), Jean-Jacques Malo (France), Stefano Maugeri (Italie), Pieter et Tjitske Van der Meulen (USA), Wayne Moore (USA), Cliff Olson (USA), Paul Paris (France), Albert Pierre Petit (France), Jean-Charles Pichon (France), Robert M. Pyle (USA), Victor Reinking (USA), Jacqueline Roumeguerre-Eberhardt (France), Russ Riddell (USA), Robert H. Ruby (USA), François de Sarre (France), Mary Schlick (USA), Pr. Franco Tassi (Italie), Bonnie West (USA), Scott White (USA/France), Dave Willingham (USA), Lila Whalawitsa (USA), Marc Yvard (France).

The author also wishes to express his gratitude to the artists who have kindly allowed reproduction of their works, in particular: Ralph Bennett, Rob Butler, Emmanuel Janssens-Casteels, Alika Lindbergh, Stefano Maugeri and Wayne Moore.

The author is particularly thankful to Jean-Michel Grandsire, publisher of *Sasquatch et le Mystère des Hommes Sauvages* (éditions Le Temps Présent, Agnières, France, 2007); to translator Paul LeBlond and author Chris Murphy, both in Canada, who were instrumental in adapting the original text for the North American public.

Contents

Translator's Note 8
Foreword 10
Introduction 12

PART I: Tracing the Mystery 18
1. Albert Ostman's Adventure 19
2. The Ape-men of Mount St. Helens 30
3. The William Roe Experience 35
4. The Bigfoot Decade, 1950–1960 37
5. Tom Slick's Expeditions 43
6. The Patterson/Gimlin Film 53
7. René Dahinden and Peter Byrne: The View from the Forest Floor 62
8. Encounters in Neah Bay and Elsewhere 75

PART II: My Pacific Travels, Insights and Experiences 80
9. The Pacific Northwest Seen from the Columbia River 81
10. The "Fabulous Beasts" Conference & Introducing "Prospector Ed" 84
11. The Columbia in the Winter 87
12. The Lesson Continues 95
13. Grand Coulee and the Great Dam 98
14. The Blessings of the Electricity Fairy 101
15. The Electric Garden 106
16. Good-bye For Now 110
17. The Dalles 115
18. The Salmon — A Priceless Gift 120
19. The Apple Trees of the Columbia 136
20. The Lonely Pedestrian 143
21. A Visit to the Museum of the Colville Confederated Tribes 148

Part III: Wild Men & Men of Science 155
22. The Sasquatch: A Few Words Before Hearing
    from the Scientists 156
23. Sasquatch and Science 162
24. A Parenthesis 165
25. Back to *Big Footprints* 169
26. The Right Choice 176
27. Phylogeny 182

28. Interlude 187
29. In Other Lands: *Gigantopithecus* or Neanderthal? 191
30. Professor Boris Porchnev 196
31. Zana, the Ogress 203
32. "Slanted, bloodshot eyes"... frequently blinking 209
33. A Few Years Earlier 214
34. The Father of Cryptozoology 219
35. The Impact of the ISC 223
36. About Some Other Unusual Folk and Creatures 228
37. The Frozen Fossil 241
38. Lingering Questions 248

Color Presentation 257

Part IV: Sasquatch and the World of Mythology 289
39. A Bridge Toward Mythology 290
40. The Literal Truth 302
41. Meeting the People of the Pacific Northwest 306
42. The Conquest: In the Shadow of Lewis and Clarke 317
43. The Indians Dispossessed 322
44. Myths: An Endless Tapestry 327
45. On the Way to Nespelem 332
46. Vi Hilbert 335
47. Through the Stone Gates 338
48. Killer Whale 344
49. Bigfoot is Everywhere 350
50. Dzonokwa and the Potlatch 357
51. The Giants 363
52. Concluding Remarks 369

Additional Photographs from Author's File 378
Bibliography 392
Photographs/Images — Credits/Copyrights 404

Appendices 409
Appendix 1 411
Appendix 2 415
Appendix 3 418

General Index 425

Translator's Note

Jean-Paul Debenat and I first met at the Fabulous Beast Conference, jointly hosted by the International Society of Cryptozoology (ISC) and the British Folklore Society at the University of Surrey in the summer of 1990. The event brought together communities of researchers who would not normally have interacted within the framework of their professional activities. ISC members were mostly natural scientists intrigued by reports of undiscovered animals, trying to make sense of incomplete and often anecdotal evidence, and trying to fit these mysterious creatures within the zoological framework. Folklorists on the other hand had more historical and cultural interests. The meeting proved most stimulating for both groups. Scientific cryptozoologists were exposed to a deeper historical perspective on wild men and their enduring presence on the margins of human society. Folklorists heard about recent reports on the creatures they knew mostly from the historical and literary record. This book is indeed an outcome of that Fabulous Beast Conference.

Through his extensive scientific and historical research on hidden animals, Bernard Heuvelmans provided a bridge between the two approaches. Jean-Paul Debenat and I are disciples of Heuvelmans, having both been drawn to cryptozoology through his works, but from different perspectives. Our relationship developed over the years at gatherings of cryptozoologists, and evolved into a close friendship through visits to each other's homes in France and Canada. I was flattered and responded enthusiastically when Jean-Paul suggested that I might translate *Sasquatch: cryptozoologie et mythologie* for a North American readership.

There is an Italian aphorism, *Traddutore, tradditore*, a translator is a traitor. I hope I have not betrayed my friend in adapting his book to a different language and a new audience. Introducing a French audience to the sasquatch phenomenon in its physical, cultural and mythological environment requires an approach that may not be ideal for readers already familiar with the Pacific Northwest. Nevertheless, although I eliminated or modified, with the author's permission, some cultural references that would not resonate with a North American readership, I have kept as closely to the text as possible, striving to keep alive the freshness of discovery that is one of the main strengths of Jean-Paul's work.

One advantage of performing a translation is that one must read in depth, checking and remembering details. Chris Murphy's encyclopedic knowledge was of great assistance when it came to ensuring the accuracy of details of bigfoot observations. I enjoyed reading about Pacific Northwest native mythology, modern theories of human ancestry, the geology and history of northern Washington State, the mighty Columbia River, Ed Fusch's inventive idiosyncrasies, and where sasquatch fits in all this. This book is more than just about sasquatch/bigfoot and its environment. It offers a broad and deep introduction to the question of wild man and its relation to modern humans through scientific as well as mythological perspectives.

— PAUL H. LEBLOND

Foreword

When David Hancock showed me Jean-Paul Debenat's French version of this volume in November 2007, I was both surprised and excited. I had met Jean-Paul at one of Ray Crowe's conferences in 2002, but had not heard or seen any more of him. Ray took a photo of Jean-Paul and me during the conference, which now and then popped up in my files, so he did come to mind occasionally.

Now, I had a large book in my hand authored by the very nice fellow I had met—thus the surprise. As I flipped through the pages, I became very excited; this appeared to be a great book, on a subject near and dear to my heart, but unfortunately it was in French. I tried to chip away at some of the material, using my Grade 12 French lessons, but such had faded to the point where my attempt was futile.

David asked me about publishing an English version of the work, so I took it home with me thinking that at least I could study the images and perhaps roughly translate some of the captions. Again all was to little avail. Over the next few weeks, I must have thumbed that book 50 times—I was dying to know what was in there! This created a bit of a dilemma for me. Here I had this vast volume of knowledge, but I could not access it—"Water, water, everywhere, but nowhere a drop to drink."

With David's decision to publish an English version (although essentially sight unseen), I was greatly relieved. Paul LeBlond had agreed to do the translation, so I knew I would be in good hands on what became a rather remarkable journey.

First off, the work is extremely well written—literature is Jean-Paul's specialty, so this was expected. However, we don't often see an author of his literary caliber apply his skills to a documentary, especially one with a fair amount of scientific content. We are therefore treated to "first-class accommodations" as we travel through his experiences, knowledge gathering, thoughts and insights. Indeed, I have not seen a book on the sasquatch/bigfoot subject that provides so much impact—when Jean-Paul explains something one virtually lives it, as opposed to simply reading interesting information.

It is very difficult to summarize this work; there are numerous ports of call. Many take us on side trips into intellectual regions not often visited by sasquatch/bigfoot enthusiasts, whatever their various walks of life. One might even wonder at times, "What are we doing here?" However, all travels have a bearing on the overall theme of the work. In short, this book is a true learning experience.

Generally speaking, Jean-Paul examines the sasquatch/bigfoot mystery from both the "white man's" perspective and that of Native Americans. We experience strong connections and vivid contrasts. Although the mystery remains unresolved, we leave with a far greater understanding of what is and could be involved in this age-old phenomenon.

I need to mention that this book differs from Jean-Paul's French version. The main difference (other than formatting and some rearrangements of material) lies in the provision of alternate and additional photographs and images. It was reasoned by David Hancock and myself that Jean-Paul's excellent tome could be suitably enhanced with the material we had (or could arrange to obtain), and other photos/images available from Jean-Paul himself. The revised work therefore provides both an impressive visual and literary experience.

— Christopher L. Murphy

Introduction

This volume is the result of a ten-year effort on my part to come to grips with the subject of "wild men." This expression sometimes describes our prehistoric ancestors, but also includes so-called primitive tribes, as well as relict hominids, and other nebulous beings—for example, the wild men of medieval art and folklore. "Wild man" is thus a vaguely defined term. In this book, it will mainly refer to the hairy giant of the Pacific Northwest, without forgetting its potential links with other categories of wild men.

My interest in wild men stems back to my teenage years, when I read Vercors' novel *Les Animaux Dénaturés*, 1952.[1] Some passages left a lasting impression on my mind. Take, for example, the following dialogue between two of the main protagonists:

> If you are not an orthogenist, *[a believer in directed evolution]* what are you then?"
>
> Nothing. **I am open.** I think orthogenesis is a mystique and, along with Darwin, that natural selection plays a major role. However, I also believe that it is not the only player; that evolution is the result of complex factors, both internal and external—of all kinds of interactions. **I don't think that evolution will ever be reduced to a simple single cause. I believe that those who believe that are dunces.**

These few lines set me thinking. For many years, busy with other concerns, I paid little attention to the wild man. Yet, it lived within me in the form of Gargantua and Pantagruel, figures familiar to me through my father's admiration for Rabelais' work.

The wild man now and then made an appearance, here as a face

---

1. Vercors: pseudonym of Jean Marcel Bruller, French novelist (1902–1991). *Les Animaux Dénaturés* relates the discovery of a missing link between man and ape in the jungle of New Guinea. Translated as *You Shall Know Them*, Little, Brown and Company, Boston, 1953. Adapted to the screen by Gordon Douglas as *Skulduggery* (1970) with Burt Reynolds in the leading role of anthropologist Douglas Temple.

carved on a church pew, there as a feminine figure appearing in woodwork at the inn, or as an actor in some folkloric event, such as at Prats-de-Mollo (Pyrenées Orientales), where the bear and the wild man are seen as close cousins.[2]

Through my studies of the English language, I became aware of the importance given to the Green Man in Great Britain, where he often appears carved in the stone of churches, a pagan symbol that reached its fullest expression towards the end of the Middle Ages and the beginning of the Renaissance (14th–15th centuries). Like Robin Hood, he stands for the untamed spirit of the forest. In French, he was called " Head of Leaves" *(Tête de feuilles)* or "Mask of Leaves" *(Masque feuilles)* because his head, and sometimes his entire body, were covered with leaves. Some choose to see in those leaves an imitation of the coarse hair that covered the wild man's body. Further, one might imagine that the colorful costume of Arlequin, of the Italian *Comedia dell'Arte*, might represent the leaves, which themselves stand for the rough pelt of the relic hominid.

My studies of comparative literature led me to numerous libraries and to the Ecole Pratique des Hautes Etudes, in Paris, where I sat in on Antoine Faivre's lectures.[3] By chance, I ran into Jean-Charles Pichon, mythologist, novelist and scenarist. We became fast friends and, guided by his impressive erudition, I discovered the works of Mircea Eliade, Roger Caillois, Carl Gustav Jung, Georges Dumézil, Robert Graves, Joseph Campbell, as well as those of countless novelists, prophets, theologians and storytellers, past, present and future.

At that time, my friend and colleague Lauric Guillaud was studying that genre of Anglo-American fiction known as " lost-world literature." There are literally hundreds of works that have followed in the footsteps of Conan Doyle's 1912 *Lost World*, where Professor Challenger discovers dinosaurs and a race of prehistoric

---

2. Prats-de-Mollo is a small medieval town in the French Pyrenees where a traditional carnival pageant features the bears of local mountains and people masked as bears.

3. Antoine Faivre, (1934–) is a prominent French scholar of esoteric knowledge, author of numerous studies on sects, secret societies and little known but influential schools of thought.

men on an isolated plateau in South America. This prolific genre came to life just one year after J. H. Rosny's *La Guerre du Feu*.[4]

Lauric Guillaud shared with me the results of his research. Following an exchange of letters with author and zoologist Bernard Heuvelmans, Lauric brought me along to a meeting with Heuvelmans at his house, near Le Bugue, in the Dordogne, a cradle of European prehistory.[5]

The events of that day are etched in my memory. We were sitting in our host's office-library, a vast elongated room crowned by the ceiling beams of what used to be a barn or an attic. Bernard answered our questions in a calm voice, with a slight Belgian accent. The walls were filled with books, a veritable treasure trove of information. Bernard showed no sign of tiring of our questioning. Suddenly, out of the small window on my right, I thought I saw the shape of a monkey. A moment later, there it was again.

"Ah!," said Bernard, smiling, "You have just glimpsed a howler monkey," as if there was nothing unusual about that!

Seeing our surprise, Bernard led us downstairs. In the yard, there where a few patches of snow—it was February. Bernard whistled. A small female leaped out of the hedge and came to nestle in his arms. There were a few other specimens here and there on the property. I voiced my surprise: how could these Amazonian monkeys, in spite of all that I had read about them, survive a winter in Perigord?

Bernard answered that these monkeys had obviously not read the same books. A few minutes later, we raised a glass to the health of these charming creatures.

Bernard became our friend. I became a member of the International Society of Cryptozoology, of which he was the president. I contributed some articles and participated—sometimes as a speak-

---

4. *La Guerre du Feu* (*The Quest for Fire*) first published 1911, describes the life of a prehistoric tribe. Adapted for the screen by French director Jean-Jacques Annaud under the same title in 1981.

5. Le Bugue, small village in the French province of Dordogne, near the caves of Lascaux — famous for their prehistoric wall paintings.

er—in their meetings, particularly one in Guildford (UK) which I attended in Bernard's company.

In Europe, I met cryptozoological researchers specializing in a variety of disciplines—genetics, paleontology, archaeology—some coming from as far as Australia, Canada or China. A great part of this book consists of memorable interactions with researchers, complemented by more bookish encounters. The scenery, the masks, the animals, the books, the people: they all have their message. It is not possible for me to reduce them to purely abstract notions: let's welcome them fully to the ample room they occupy here. They are not mere signposts along the path that I invite you to follow; they are the path itself.

In 1994, the University of Nantes granted me a sabbatical leave, which allowed me to continue my research *in situ*. I left to explore the splendor of the Pacific Northwest, from majestic Mount Rainier, the Cascade mountains, to the desert of eastern Washington, near Idaho. In this wild and spectacular environment, the snickering and knowing smiles of urban academia were clearly out of place. In this region, the word "sasquatch" is a source of curiosity among white Americans, of inspiration among Natives.

I had to spend many years selecting among the countless accounts of enquiries, communications, and works I accumulated, to document my interest in wild men. Often, instead of writing, I would plunge into a recently discovered new book; I would lose myself in the pages describing sceneries of today and yesteryears; I could hardly tear myself away from a land where I had begun to take root. The manuscript progressed slowly. In the end, the third part of the book allowed me to adopt a personal approach, which I actually borrowed from Geronimo:

> During the sessions when he narrated, Geronimo freely passed from one event to another in his life, in a characteristically Indian fashion, which consists in saying only what seems important to the narrator, and to say it in the manner and order which he thinks most appropriate. *(Memoirs of Geronimo.)*

On second thought, I understand why writing this work went so slowly: one day, I realized that I was the spectator of a research, some would say a quest, by actors from North America to China, from Pakistan to Australia—I was also one of those actors, by way of the talks that I presented in Europe, and of my exploratory visits to the Pacific Northwest.

I experienced the pleasure of delving into a scientific adventure; it taught me the patience required to come to understand a complex situation. I also experienced the joy of wading into a mythological adventure; my relation with Jean-Charles Pichon, as an author and a friend, had prepared me for it. Suddenly, I was face to face with myth in its very evolution. I had become a participant. Given such an opportunity, how could anyone stand back and reject with a clear conscience the developing myth? And, watching the myth develop, how could I not want to tell others?

Believe me! Curiosity, fascination and intuition will certainly prevent me from concluding prematurely my quest for the wild man.

I have attempted to put some order in the stories, accounts of various works, and visual documents that have nourished my research.

In the first part of this book, I shall relate encounters with sasquatches which took place from the 1920s on. In the next part we will discover the Pacific Northwest, setting the scene from Northern California to British Columbia, through Oregon, Washington and Idaho. We will discover the geological and human aspects of the region, with the Columbia River as a reference point, dividing feature, and entry into the description of different life styles.

The third part of the book presents the sasquatch as scientists describe it in their writings. In the spirit of cryptozoological research, we will follow the methods of researchers in the field, learned or otherwise, led by the certainty that there exist creatures that science has not yet classified. Hypotheses follow each other, points of view clash, fiction draws inspiration from the latest theories.

The fourth and final part addresses the traditional aboriginal context, particularly the dominant figures of the bestiary of Native culture. By then, it will have become obvious to the reader that it is not possible in practice to separate the scientific approach from the traditional aspects of the wild man. Nevertheless, the last part of the book emphasizes the tales and legends of the shamanic universe, where there is no real distinction between the real world and the realm of the sacred.

At that point, the narration becomes more personal, expressed in the first person. The path followed is clearly my own, but it is also presented as an invitation to the reader to follow his or her own itinerary and to develop their intimate reflection on our origins and the potential for the existence of other members of our race.

Before venturing any further, readers should note that they are about to discover a variety of theories, as well as some unusual people and unfamiliar locations. They are cordially invited to enjoy all these encounters. In spite of my efforts, the names and the words may sound unusual and the vocabulary rather specialized. The words are intimately linked to the entities, objects and settings encountered in the search for the wild man. Take your time, dear reader, and slowly follow its footsteps; do not hesitate even to retrace **its** steps.

Part I

Tracing the Mystery

> " *until chance leads us to discover some other species similar to ours, since there is no reason that there couldn't exist some such in hitherto unknown regions. This animal is so much like us that naturalists have called it wild man, or man of the woods."*
>
> JULIEN OFFRAY DE LA METTRIE
> *L'homme-machine* (1747)[1]

---

1. Julien Offray de La Mettrie (1709–1751), French physician and philosopher, the earliest of the materialist writers of the Enlightenment. He has been hailed as a founder of cognitive science.

Chapter 1

Albert Ostman's Adventure

I have always followed logging and construction work. This time I had worked for over one year on a construction job and thought a good vacation was in order. B.C. is famous for lost gold mines. One is supposed to be at the head of Toba Inlet—why not look for this mine and have a vacation at the same time? I took the Union Steamship boat to Lund, B.C. From there I hired an old Indian to take me to the head of Toba Inlet.

Thus begins Albert Ostman's story as he meticulously recorded it in 1954, thirty years after the fact, prompted by John Green, then a journalist in Agassiz, a small town about 60 miles (100 km) east of Vancouver, B.C.

The Indian spoke of a white man of long ago, a prospector returning with a bag of gold from an abandoned mine on Toba River, who was spending his fortune in the saloons. One day, however, he did not return and the rumor was that he had been killed by a sasquatch.

Albert had never heard of the sasquatch. The Indian explained that it was a hairy being, human nevertheless, and very tall, who lived in the mountains. His uncle had seen 25-inch-long (65 cm) footprints and an old Indian had seen an eight-foot-tall (2.44 m) sasquatch. Albert's reaction was that he did not believe those stories of mountain giants. Perhaps they might have existed some thousands of years ago, but they had long since vanished. The Indian conceded that there were indeed very few of them left, but that they were nevertheless real.

Albert set up camp at the mouth of a stream while the Indian waited for high tide to return to Lund. Leaving Albert around 7:00

Albert Ostman

p.m., after dinner, he promised to come and pick him up in three weeks.

Albert had with him a Winchester 30-30 and a homemade prospector's pickaxe, pick at one end, axe at the other, held in a leather holster on his belt, next to his knife. He carried boxes of sugar, salt and matches, cans of food, lard, a bag of beans, four pounds (1.8 kg) of dried prunes, six packages of macaroni, three pounds of pancake flour, cheese, six packages of rye biscuits, three tins of chewing tobacco (snuff), a waterproof can of butter, two cans of milk and two boxes of shells. He buried a box of biscuits on the spot as a reserve for his return. All these details, Albert provided to John Green from a shopping list, which he had kept.

The following morning, he rolled up his sleeping bag and tied it and his groundsheet to his backpack, along with his frying pan and aluminum pot. The empty tins would be used for cooking.

He started off early, with at least eighty pounds (36 kg) on his back, not including the gun. He then began a steady and sometimes strenuous climb, ending up in the evening at about the 1,000-foot (300-m) level, high enough to enjoy a panoramic view of the islands and the strait, dotted with fishing boats and tugs pulling log booms.

In the days that followed, Albert continued toward the northeast—climbing, crossing mountain passes, turning back when stuck in dead-end ravines. His prospecting remained fruitless, but having killed a deer, he could enjoy fresh venison.

After a week, he came upon an ideal site and decided to set up camp there. He cut up a bed of branches which he laid out under two cedars growing near a cliff. He stuck a pole into a crack in the rocks to hang his bags. With flat stones, he built a small oven to cook a grouse, which he had just shot.

In the morning, after a sound sleep on his bed of branches, he noticed that some of his utensils had been moved. However, nothing was missing.

Albert wasn't concerned. He continued prospecting, without success. Returning to his camp, he roasted a squirrel, opened up a can of peas, and with dead fir branches stoked his fire, which provided ample light and heat.

Before bedding for the night, he loaded his rifle, suspecting that a porcupine might be prowling nearby. He hid his shoes at the bottom of his sleeping bag, knowing that these animals love leather.

In the morning, he noticed that his backpack, hanging on the pole, had been upset. A bag of prunes was missing, as was the flour, but the salt, a favorite of porcupines, was still there. The guilty party must have been some other animal. Too bad, thought Albert, a porcupine stew would have been just fine.

Three nights in a row, the nocturnal visitor left clues of its passage. In the morning Albert sat on a prominent rock and scrutinized the area without discovering a trace. He considered moving his camp, but decided not to. There were too many advantages to the chosen site, especially the pair of large cedars that provided shelter from the rain—the sky was clouding up.

> I took special notice of how everything was arranged. I closed my pack sack, I did not undress, I only took off my shoes, put them in the bottom of my sleeping bag. I drove my prospecting pick into one of the cypress trees so I could reach it from my bed. I also put the rifle alongside me, inside my sleeping bag. I fully intended to stay awake all night to find out who my visitor was, but I must have fallen asleep. I was awakened by something picking me up.

Half asleep, Albert wondered where he might be. He realized that he was still in his sleeping bag—carried away by an avalanche, perhaps. But where from? Then he got the feeling that he was being thrown onto the back of a horse. But it was not a horse; it was a being that walked upright!

Albert tried in vain to draw his knife out of its sheath, but he managed to grab his gun. He could feel his packsack and the tin cans it contained painfully poking his back.

After about an hour, he guessed that his kidnapper was climbing a steep slope. He could hear it breathing heavily.

Scrunched up at the bottom of his sleeping bag, with one of his upturned hobnailed boots hurting his feet, Albert was in a painful position. Sometimes the bag hit the ground and he ended up being dragged along. He suffered from horrible cramps. Finally, after about three hours, he was dropped on the ground. He heard voices speaking an incomprehensible language. The ground was sloped and he rolled for a few moments. He tried to get out of the bag, but his legs were asleep. He rubbed them vigorously.

Four silhouettes stood around him, in the dark. Without being able to see clearly, he realized that he was now the prisoner of those mountain giants mentioned by the Indian.

The four giants continued their conversation as the day broke. He could now see two adults and two smaller beings, completely hairy and without clothes—a family (of Sasquatches?): father, mother, a boy and a girl.

The children seemed afraid. The mother appeared unhappy with her "husband's" catch; he gestured as if to justify it and to explain what he had in mind. The four of them then left.

Albert examined his surroundings: a boxed-in meadow, a few acres in area, with a V-shaped entrance in the east. He dropped his possessions under a couple of small cedars and took stock. Canned foods, coffee and butter were still there, but the prunes and macaroni had disappeared. The gun was loaded and he had six shells in reserve. His pickaxe was gone.

There didn't seem to be any dry wood around. The grass was green; there must be some water nearby. Albert dropped his coffee grounds in a rag and went in search of a spring.

The children were watching him from behind a juniper bush. Albert filled up his tin can in a stream. Returning, he saw the boy examining his possessions, but without touching them. He also discovered under an overhang a platform covered with dried moss. Lying on top of that "carpet" were rough blankets woven from narrow strips of cedar bark mixed in with dry moss. They looked warm and practical.

The following morning, Albert prepared to leave. He rolled up his sleeping bag, stuffing it into his backpack with the remaining food cans. Loading his gun, he walked toward the opening of the canyon. However, the "father" stood up, hands stretched out, ready to push him back, saying *Soka, soka* or something like that. Albert walked back about 60 feet (18 m), but hesitated to shoot. He was not sure he could actually kill that giant and feared that merely wounding him might provoke his fury. He figured that there might be a better way to escape.

The following day, the female sasquatch disappeared and returned around 4:00 p.m. with an armful of grass and twigs of all kinds—from fir boughs to hemlock, as well as hazelnuts. The son offered Albert a long-rooted, sweet-tasting plant. In exchange, Albert offered his snuff box with a few grains still in it—he had a plan. If the son took the box to his father, he might have a taste of snuff and like it. The son ventured a careful taste and showed the box to his father, who licked the remaining tobacco with apparent relish.

Albert thought things over; perhaps if he now befriended the father, he might get him to eat a whole box of snuff—enough to kill him, no doubt. Then he (Albert) would be able to escape without having to shoot anyone.

He had by now been a prisoner for six days and he noted with some satisfaction that he was gradually befriending his guardians.

He'd had ample opportunity to observe them. Their agility was fascinating. They climbed up cliffs with the speed and assurance of a mountain goat. To sit, they bent their knees and went straight down. They stood up without help from their hands. All their activities seemed oriented towards a specific goal—namely seeking roots and plants to feed on. Albert had never seen them eating meat or engaging in any cooking activity. Outside foraging periods, the parents rested. The children were always playing. The son grabbed at his feet, arched his back, and leapt forward, each time jumping 20 to 25 feet (6 to 7.6 m) ahead.

Each day the Sasquatches came a little closer to Albert. It was his only solace. What did they want from him? They did not harm him. They merely observed him, especially when he took a pinch of tobacco, which he then methodically chewed.

Finally, one morning as he was preparing his breakfast, the father and the son were sitting about 10 feet (3 m) away from him. Albert thought that they might have been attracted by the smell of the coffee. He opened his tobacco tin. The father approached, grabbed it, ate the contents at one go, and licked the box.

Soon he started rolling his eyes. He grabbed the coffee pot, swallowed the rest of the coffee and the grounds at the bottom; then he rolled on the ground squealing like a pig in an abattoir. Albert grabbed his gun; the father ran towards the stream. The son ran towards his mother, now whining.

It's now or never! Albert grabbed his pack and fled, the mother after him. He ran through the V-shaped gap, turned around and fired a shot over her head. She immediately turned back.

Albert reloaded and kept running. He covered two to three miles (3.2 to 4.8 km) in record time, stopping in a ravine. Two mountainous crests blocked the view. He climbed up the first to see if the Sasquatches were pursuing him and to try to see the coast. At the top, he saw Mt. Baker. He was reassured and reoriented. No one was pursuing him.

Tired and hungry, Albert opened his last can of corned beef. He didn't dare light a fire; he rested for two hours before continuing. Around three in the afternoon, he began to descend toward a river. He killed a fat grouse. Hiding between two rocks, he roasted the grouse, brewed a coffee and enjoyed a "real meal." He now felt safe and spread his sleeping bag below a spruce.

The following morning, he woke up stiff, his stomach in a knot, his feet sore, his strength drained. He set out, his legs shaking, and was forced to stop every 100 yards or so. It took him six hours to reach the second crest hidden in the mist.

After a two-hour trek down the other side, Albert reached a dense forest; from time to time, he heard the sound of a motor. He knew he was saved: there were loggers nearby. He heard a tree fall and walked out of the bushes. Men surrounded him. One of them asked, "You look like a wild man of the woods! Where are you coming from?"

Albert explained that he was prospecting for gold and that he got lost wandering in the mountains. He was taken in a truck to the logging camp, on the coast. Fed, washed and shaved, he rested there for a day before going back to Vancouver by boat.

Here's how Albert ended his story:

That was my last prospecting trip, and my only experience with what is known as Sasquatches. I know that in 1924 there were four Sasquatches living, it might be only two now. The old man and the old lady might be dead by this time.

Albert Ostman waited 30 years before telling his story to journalist John Green. Green later paid Ostman another visit, on August 20, 1957, in the company of A.M. Naismith, a police magistrate, who wrote up an affidavit in which he states in part:

> I found Mr. Ostman to be a man of sixty-four years of age; in full possession of his mental faculties; of pleasant manner, and with a good sense of humor. I questioned Mr. Ostman thoroughly in reference to the story given by Mr. Green. I cross-examined him and used every means to endeavor to find a flaw in either his personality or his story, but could find neither.

The world is full of great storytellers, especially far from cities—in the country, the deserts, the mountains and the oceans. Hunters, fishermen, gold seekers, foresters and farmers relate stories imbued with fantasy, humor and mystery. One is readily taken in, and city dwellers appreciate these tales as if they had too long been deprived of their magic. A good storyteller uses his voice, his looks, mimes the tale with his body, sometimes using special effects. Novelist and storyteller Pierre-Jackez Hélias[1] related one evening that his grandfather would take him as a child to gather twigs, ferns and dried herbs, which he held in his "story-bag." While telling the story, his grandfather drew from the bag and discretely threw a fistful into the fire—twigs for crackling, dried grass for bright sparkling. Light and sound effects

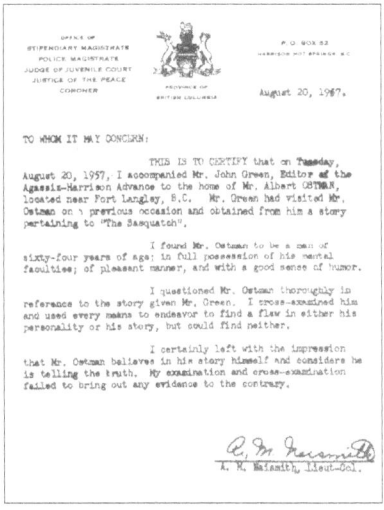

A.M. Naismith's affidavit of his interview with Albert Ostman.

---

1. Pierre-Jackez Hélias (1914–1995), Breton writer famous for his autobiographical work Le Cheval d'orgueil (1975) [Horse of Pride]. P.-J. Hélias was a gifted storyteller who occasionally gave public performances.

punctuated the story. One of the many, and perhaps one of the main charms of a story lies in the evocative precision of the descriptions: scenery, flora and fauna, climate, sounds and smells. Places and times come to life and acquire a tangible flavor.

Albert Ostman's tale takes us into the very heart of the wilderness of the Pacific Northwest. His description of the scenery is flawless, save the visibility of Mount Baker (he probably mistook another mountain for it). Even today, one meets solitary prospectors, similarly outfitted, still searching for gold. They carry clothing, tools, kitchenware, provisions, arms and a sleeping bag. What a marvelous story—the adventures of a man who left his town or village to prospect deep in the vast forest, fording a stream, climbing a rocky hillside, kneeling over the tracks of some animal, shooting game, picking a sheltered site where to light a fire, suspending his supplies out of reach of scavengers. We hear with him the wind in the leaves, the cracking branches, the babble of a brook. We stand in admiration of the starry sky, or in anxiety at the arrival of dark rain-laden clouds. We see the shadow of the giant pines and, just after falling asleep, we are startled like him by the rustle of an animal in the bushes. We imagine the shadow of a porcupine magnified like an antedeluvian monster before disappearing in the undergrowth.

There is no doubt that even a straightforward report, as told by one familiar with the setting—in this case the coniferous forest of British Columbia—may sound like a tall tale. It soon becomes one when it speaks of unusual creatures such as hairy giants.

To stick to the facts, here is a precise description of those creatures according to Albert Ostman, for whom there was no doubt that these creatures are human:

The mother was perhaps between 40 and 70 years of age. She was 7.5 feet (2.3 m) tall and weighed between 500 and 600 pounds (227 and 272 kg). She had pendulous breasts, wide hips and waddled like a goose.

The father was about 8.5 feet (260 cm) tall. He was barrel chested, with powerful shoulders, enormous biceps and forearms, and had a hump on his back. His forearms were well proportioned, but rather longer than usual. His hands were quite wide and his palms concave like a ladle; the fingernails looked like gouges. His feet were immense, with pads, like a dog's paws. The big toe was large and powerful, built for climbing steep cliffs.

Albert Ostman (left) being interviewed by John Green in 1957.

These creatures were covered all over with hair, except in the palms of their hands and the upper part of their noses. Albert Ostman did not see any ears, hidden perhaps by the long hair which reached 10 inches (25 cm), and even longer in the female.

The attitude of these giants towards their "guest" is summed up in a single word: benevolence.

In the wake of such an extraordinary and detailed story, it is wise to hear what the experts have to say.

Let us begin with John Green, a professional journalist well known for his honesty and experience. He has devoted more than 50 years to the study of the sasquatch, presenting some 2,000 reports, which he gathered in a series of books. Today, he lives in Harrison Hot Springs, B.C., 60 miles (100 km) east of Vancouver. That's where I met him.

According to John Green, today one would laugh at such a tale, suspecting its author to have been inspired by similar stories published in books or magazines. However, other than the odd newspaper report, very little information on what we now call "sasquatch" was available in or before 1924. It was not until the 1960s that reports and studies of these creatures would begin to be published. In further support of Ostman's account, Green points out that Ostman's description of the creatures' appearance has been supported and confirmed by later witnesses.

Ivan T. Sanderson, explorer, naturalist and author of popular works of natural history, paid Ostman a visit in British Columbia and described the encounter in his 1961 volume *Abominable Snowmen: Legend come to Life.*

Drawing by Ivan Sanderson under Ostman's direction depicting the adult male sasquatch that Ostman encountered.

To Sanderson, Ostman was clearly more than just a rustic woodsman. He had some bookish knowledge, spoke two languages (Ostman was originally Swedish) traveled extensively and was interested in world affairs. He was also acutely aware of the power of ridicule. Comfortably retired in his modest home, he was surrounded by friends and did not seek shrill publicity. Later in life, in contrast to his feelings when younger, he no longer feared being laughed at. He was deeply moved by his adventure and strove to explain the mystery of his capture. In this connection, Sanderson stated, "Mr. Ostman is, in fact, sick and tired of skeptics."

Sanderson focused upon a number of points, the most important being the structure and relative complexity of the language used by the Sasquatches. They carried extensive verbal exchanges. Sanderson spoke of a "delightful expression" used by the son when pointing with his finger (to his father) at an unfamiliar object with strange contents—the tobacco tin. *"Oook,"* said the son, in the manner of a cartoon caveman.

Sanderson also commented on the feeding habits of the family of vegetarian giants. The mother washed bundles of branches and leaves before stacking them. Sanderson was not surprised, but John Napier was. The latter, a renowned primatologist, noted that the sasquatch family was equivalent in weight to five gorillas, or to 14 average-sized humans. Grasses and fir boughs could hardly provide enough nutrition for such a biomass. Napier figured that such large creatures must consume a very large amount of low energy food, as does another similar weighty vegetarian, the mountain gorilla, which eats loads of celery, bark, bamboo, nettles and a variety of roots. Just like the gorilla, the sasquatch should be spending most of its waking hours foraging for food—especially since the forests surrounding the Toba River estuary yield poor quality food, typical of the conifers that grow in this region. In addition, in its tropical environment, the gorilla enjoys a year-round supply of roots and foliage.

Napier concluded that, "Albert Ostman's story fails to convince me primarily on the grounds of the limited food resources available." However, he found "no obvious inconsistencies" in Ostman's report.

Sanderson also considered with interest the description of the Sasquatches' shelter, which was of the kind used by Stone Age people. On a platform, about 10 feet deep (3 m) and 30 feet wide (10 m), the creatures had spread a bed of branches, moss and dried grass. To sleep, they covered themselves with bags of woven bark, also filled with moss and dried grass. However, it needs to be said that John Green has stated that no other report mentions such "mossy" blankets.

Sanderson speculated on these observations, thinking that perhaps before the use of stone or bone tools and the discovery of fire, there might have been a stage when food gathering was accompanied by weaving techniques. This skill requires great dexterity, even to produce rather coarse results, but Sanderson found this hypothesis neither unreasonable nor illogical.

Overall, both Napier and Sanderson found that the idea of a vegetarian sasquatch, speaking a primitive language and unaware of how to make fire, clothes or tools, was not implausible.

Chapter 2

The Ape-men of Mount St. Helens

In 1964, Roger Patterson (whom we will discuss later) went to Willow Creek to enquire among the locals if anyone had seen sasquatch tracks. He met with a number of other investigators, and an article published in the Portland-based *Oregonian* newspaper on July 13, 1924 was brought to his attention. It told the story of five miners whose cabin, which was near Mount St. Helens, Washington, was attacked by what were termed at the time, *mountain devils*—long spoken of by the Indians.

The miners were working a claim on the Muddy River, an affluent of the Lewis River, about 7.5 miles (12 km) from Spirit Lake. Over the previous few years, they had noticed large human-like footprints, but had never seen what had made them. When the newspaper article was published, most people were skeptical of the miners' story—except the Indians, who avoided the Spirit Lake region because they feared the "spirit" that they believed inhabited the area.

Mount St. Helens with Spirit Lake in the foreground (left) as it appears today. The volcanic mountain erupted in 1980 and laid to waste the forest in all directions. See the color section for a spectacular view of the summit. (Image from Google Earth © 2008, Digital Globe; TerraMetrics.)

In 1966, Patterson traced down Fred Beck, one of the three miners still alive. Beck agreed to an interview by Patterson and it was published by John Green in his book *On the Track of the Sasquatch*.[1] Here, in part, is what Beck related (selected quotations):

> We weren't scared, but my old man [his father-in-law] believed in them, and he always carried a loaded gun with him when he went to the stream. One day, coming back to the cabin, he saw a creature looking at us from behind a tree, about a hundred yards away. He raised his gun and shot it three times.
>
> "Don't run Fred," he said. "Don't run. He won't go far. I put three shots through that fool's head. He won't go far."
>
> We walked up the ridge and looked down and there he was going, jumping fourteen feet at the time. The old man shot twice again.
>
> "My God! I don't understand it," he said, "how that fella can get away with them slugs in his head. And I hit him with the other two shots, too."
>
> We then went back to the log cabin. We had built a solid roof of good-sized pine logs for rafters and two-inch pine shakes to hold the snow, with pine branches for chinking.
>
> After dinner, we went to bed. The attack began with a pelting of big rocks. Then them buggers lept on the roof, and knocked the chinking out onto my father-in-law's chest. There was an axe there; a hairy hand and arm reached in and grabbed it. The old man grabbed the axe and turned it so that it wouldn't go through the chinking hole between the logs, and then he shot on it, right along the axe handle and

---

1. First published by Cheam Publishing, Agassiz, B.C., 1969. Republished by Hancock House Publishers, Surrey, B.C., third printing, 1994.

An elderly Fred Beck poses for John Green, looking into the distance, legs crossed, his finger on the trigger of the large caliber rifle with which he shot at the "mountain devil." Did he tell the truth? Or did he, with his companions, merely add another page to the legend of what is now known as Ape Canyon?

it [the creature] let go of it. And then the fun started! It lasted most of the night.

My God! They made a noise. Sounded like a bunch of horses running around up there on the roof. How many were there? Impossible to know. There were no windows in the log cabin. The only part of these creatures that we saw was the hairy arm. There might have been only two of them.

The next day we found tracks. Anywhere there was any sand on the rocks we found tracks of them. I had to return to the mine to gather some tools to take home to Kelso. I

Immediately after the incident, investigators went to the scene. This 1924 photo shows the investigators standing by the besieged cabin. From left to right: Burt Hammerstrom, freelance writer; Bill Welch, forest guard at the Spirit Lake Ranger Station; Frank (Slim) Lynch, Seattle newsman; and Jim Huffman, forest ranger for the Spirit Lake district.

Fred Beck (left) and Marion Smith (one of the other miners). They are re-enacting the incident for the investigators.

took my gun with me. I saw one of those fellows about a couple of hundred yards down the ridge and I shot him in the back, three shots, and I could hear the bullets hit him and see the fur fly on his back. He fell into a precipice and into the canyon.

In the afternoon, the sun came out, we went down to the bottom of the canyon. But the torrent down there was so strong that it washed everything away.

According to Fred Beck, the creatures were at least eight feet (2.4 m) tall. They looked like a man, narrow at the waist, with broad chest and shoulders, bull necked and with large flat ears similar to ours. They had a flat nose on a hairy face. Their eyes, though poorly seen, did not look human. The depth of the footprints suggested a great weight, at least 600 to 800 hundred pounds (270 to 360 kg) according to Beck.

Commenting on the interview, John Green concluded that there really was an attack. That if such creatures do exist, there would be no need to embellish reality; in other words, why would the miners have invented such a dramatic story, improbable as it might seem, which they repeated without change to the end of their lives?

At the time, most of the residents of Kelso, where Fred Beck and his father-in-law lived, were skeptical of the miners' story. The only people who believed them were the foresters, who were acquainted with Fred's father-in-law, whom they respected as a seasoned and courageous hunter.

The canyon into which the creature fell. It is now called Ape Canyon as a result of the miners' experience. Mount St. Helens, before it erupted, is faintly seen in the background.

Chapter 3

The William Roe Experience

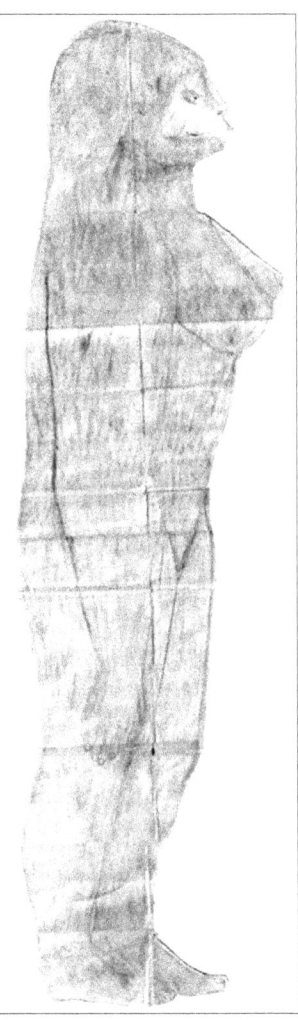

Drawing by William Roe's daughter, Myrtle, created under his direction depicting the sasquatch he saw.

Often, especially in Canada, witnesses prepare a sworn affidavit in front of a public official. Such official statements acquire enhanced credibility. Thus did William Roe submit a report of a surprising encounter near Tête Jaune Cache in a sworn statement to Justice of the Peace William Clark, in Alberta in 1955.

Roe had been working for a road construction company for two years when, in October of that year, he decided to explore an abandoned mine, five miles (8 km) away up Mount Mica. Experienced as a hunter and woodsman, he packed up his gun and started climbing the mountain.

At about three in the afternoon, he saw a silhouette, which he took for a grizzly, in a clearing near the mine's entrance. He soon realized that the creature, which was about six feet (1.8 m) tall, looked more like a human than a bear. He estimated its weight at 300 pounds (136 kg). Covered with brown hair that had silver tips, the creature, judging from the size of its breasts, was a female. The massive torso did not slim at the waist. The creature approached within 20 feet (6 m) of the bushes behind which Roe was hiding, crouched down and began eating leaves from a bush (drawing branches through its teeth). Its face was covered with hair, except near the mouth, nose and ears, so that it looked as much like an animal as it it did a human. Finally, the creature detected Roe's

35

View from the only motel in Tête Jaune Cache with Mount Mica in the immediate background. Very little has changed since William Roe had his remarkable experience in 1955.

odor and gave a rather comical stare when it discovered the man in the bushes. Remaining crouched, the wild creature took a few steps back, stood up, turned around and walked away. Roe thought of shooting it—the specimen would be of great scientific interest. He had heard stories about the sasquatch, the hairy giant of British Columbia Indian legends, a giant which they insisted was still around. Aiming, he then lowered his gun:

> Although I have called the creature "it" I felt now that it was a human being and I knew I would never forgive myself if I killed it.

William Roe's affidavit.

Describing the creature's walk, Roe said that it touched the ground with its heel first. A remarkable observation, commented John Napier, typical of human locomotion.

Chapter 4

The Bigfoot Decade, 1950–1960

The year (1955) in which William Roe had his encounter with the "wild man of the woods," or sasquatch, as the Salish Indians now call it, is an important milestone. It was from that time on that reports began to proliferate and be featured in the press. The newspaper articles kick-started the careers of many hunters and researchers. The word "bigfoot," became the common name for the creature in the United States in 1958,[1] and is now frequently used together with, or in place of "sasquatch." This term is actually the anglicized version of the Native word *Saskehavis*. It was created by John W. Burns, a teacher on the Chehalis Indian reservation, British Columbia, in about 1925.

For Roger Patterson, the quest that haunted him to the end of his days started by reading a 1959 article by Ivan Sanderson.[2] The events described by Sanderson took place far away from British Columbia; they were in California, at the southern limit of the Pacific Northwest.

It all happened in October 1958 at the fringe of Humboldt and Del Norte Counties, in the northwest corner of California (see map). Gerald Crew (Jerry to his friends), a road construction worker, drove a bulldozer at a work site in a vast forested zone, about 10,000 square miles (27,000 km^2) in area, that reached to the Pacific Ocean in the west, and to the Oregon border in the north. On the south side, State Highway 299 runs eastwards, crossing US Highway 99 (now Interstate I-5) 129 miles (208 km) from the ocean. Contrary to common belief, California, though densely populated in some places, is practically or completely uninhabited in other areas.

---

1. The word "bigfoot" was in common use in Northern California prior to this date. An Associated Press release in October 1958 on findings by Jerry Crew (later discussed) used the word and thereby brought it to world attention.
2. I.T. Sanderson, "The Strange Story of America's Abominable Snowmen." *True* magazine, December 1959, New York.

MAP OF CALIFORNIA

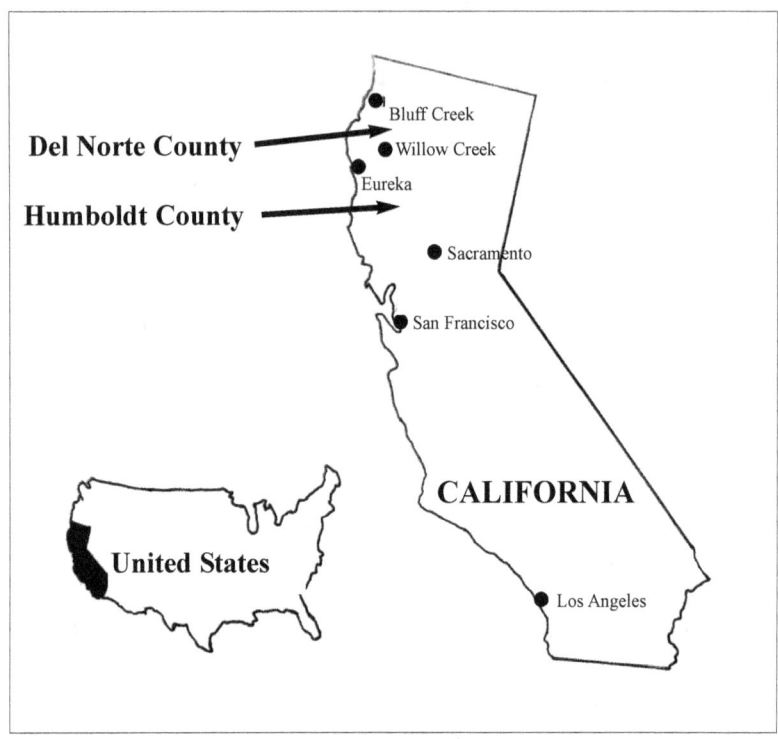

According to John Green: "The area penetrated by the new road, up the valley of Bluff Creek, was completely roofed in with closed-canopy forest and totally uninhabited."

Only a few years earlier, noted Sanderson, "a Stone Age Indian named Ishi—the last survivor of a race of men long believed extinct," had emerged from the wilderness and revealed himself to a butcher in Oroville, only about 62 miles (100 km) from the state capital, Sacramento. Jerry Crew's adventure, on the other hand, took place in an area free of any trace of civilization.

To again quote Sanderson: "The country is mountainous; though this is the understatement of the year, being to most intents and everywhere almost vertical so that you can only go up on all fours or down on your bottom. Unless you make an exaggerated and exhausting climb you cannot see more than about four square miles (10.4 km^2) of the country because you are always on the side of something going either straight up or almost straight down..."

The landscape is just as important as the creatures that inhabit it! We already came to appreciate that basic rule in earlier pages; it will surface again and again in later chapters.

Jerry Crew (d. 1993) was a native of Salyer, a village in Humboldt County. A practicing Baptist, he did not drink alcohol and had a reputation of honesty, together with calm and solid common sense. Jerry had been hired by Ray Wallace, a road construction contractor, along with other local men, including Crew's young nephew, James Crew, and two seasoned Native loggers from the nearby Hoopa Indian reservation. For the past three months, this team had been clearing away stumps and improving a logging road. On the morning of August 27, 1958 Jerry noticed a series of tracks around his bulldozer. Each print was about 17.5 inches (44.5 cm) long; successive prints were 51 inches (1.3 m) apart! The tracks started from a steep (75°) slope, they circled the bulldozer, and aimed towards the work camp.

At first, Jerry thought that this must be some kind of practical joke—although one that required considerable effort! And, how would someone making such prints avoid being noticed? Nevertheless, he had heard of similar prints being found by a road gang about one year earlier at a location eight miles (13 km) north. He showed the prints to his workmates, and some said they had also seen similar prints in the area. Only the Indian workers remained silent.

A month later, new tracks appeared, and again, three weeks after that. Sanderson remarked:

> This was continuing sub-rosa when, on October 2, the maker of the tracks appeared again on his apparently rather regular round, leaving tracks for three nights in succession and then vanishing again for about five days. This time Jerry Crew had prepared for his advent with a supply of plaster of Paris and made a series of casts of both right and left feet early one morning.

The last appearance of bigfoot occurs on the eve of Ray Wallace's return from a business trip. Alarmed by the rumors regarding the footprints, he feared that some malicious rivals were attempting to disrupt his work by scaring his employees. His brother, Wilbur, who worked with him, had examined the tracks and had provided a report on them together with other unusual events.

First, a 40-gallon (220-liter) barrel of fuel had been carried over some distance (as the tracks indicated) and was thrown down a hill. Then a heavy galvanized steel culvert disappeared and was found at the bottom of another hill a short distance away. Finally, a 280-pound (125 kg) tire was partly rolled and carried about one-quarter mile and tossed into a ravine.

Ray was quite skeptical, but large prints he found near a stream convinced him that someone was up to no good. He immediately hired a man, Ray Kerr, who wanted to work at the site and spend his spare time looking for the "culprit." Kerr brought with him a friend, Bob Breazele, a professional hunter who owned four dogs. Breazele did not work; he spent his time searching the area.

Jerry Crew with a cast he made of a footprint he found at the Bluff Creek road construction site. This photo was published in the *Humboldt Times*, October 14, 1958. The cast was about 17.5 inches (44.5 cm) long. A similar photo was used in an Associated Press release on the story, which received worldwide attention. The use of the term "bigfoot" in this release resulted in it becoming the common name for the creature in the United States.

Kerr and Breazele found more tracks, which they followed, but to no avail. Then, one night at the end of October, while they were driving down part of the newly completed road, they spotted some kind of hairy giant, crouched on the shoulder of the road. The humanoid, silhouetted by their vehicles headlights, leaped up, crossed the 20-foot-wide (6 m) road in two steps, and disappeared into the underbrush. In the days that followed, the hunters beat the bush in vain. In one of these forays, the dogs vanished, never to be seen again.

Ray Wallace was still faced with a mystery. He collected droppings believed to have been from the creature—equivalent in volume to that of a half-ton horse! He also gathered some dark-colored hair, reaching up to 10 inches (25 cm) in length. It was obtained from the bark of pine trees at a height of 6 feet (1.8 m). Was all of this the work of a practical joker? This hypothesis seemed increasingly improbable.

The *Humboldt Times* (Eureka, California) published a series of detailed articles on the "incidents" at Bluff Creek. The first included a photo of Jerry Crew holding one of his casts. Although the articles treated the subject seriously, in the town of Willow Creek Ray Wallace's crew was being ridiculed.

The Indians in the area had some interesting comments. One of them, a member of the Yurok clan, declared, "The bigfoot were run out of this country by the miners in the 1848–49 gold rush. Before that, there were quite a number of them." Another Native, a Hoopa, sourly commented, "Good Lord, have the white men finally got around to that?"

To conclude, Sanderson hypothesized the existence of a creature intermediate between man and animal, living in isolation in a vast and barely explored area rich in vegetation, water, and food in the form of berries and small game. Sanderson's long article lit a spark of curiosity in Roger Patterson, then living near Tampico (Yakima County), Washington. He quizzed the Indians of the Yakima reserve, of whom many firmly believed in the existence of "forest giants." Most hoped never to meet the creature, believing it to be an evil spirit. Others affirmed that the giants continued to wander in the local mountains, and that they were formerly said to be found everywhere in northwest America.

Roger Patterson and his now-classic book. He was the first person to compile a book about bigfoot sightings and incidents. It has undergone four printings.

Patterson, whose main income came from rodeo work, decided to check out Sanderson's story for himself. His research led to a series of expeditions of which the outcome will be discussed later. During his initial research, Patterson gathered eye-witness accounts and bigfoot-related newspaper articles. As a result of this work, he wrote and published a book (1966) entitled *Do Abominable Snowmen of America Really Exist?*[3] Patterson quoted in his book a considerable number of testimonies, including those of the Indians, to whom he listened very attentively. In the words of a member of the Yurok clan, "The Big Foot was known to my people as the traveler or the patroller. It's his job to keep nature and men in harmony."

Jerry Crew (right) with Andrew Genzoli, editor of the *Humboldt Times* (Eureka). They are examining Jerry Crew's cast. In the late 1950s, Genzoli published several other bigfoot-related articles, all of which where highly instrumental in encouraging people to "take up the search."

---

3. Franklin Press, Yakima, Washington, 1966.

Chapter 5

Tom Slick's Expeditions

The Indians' comments were undoubtedly an encouragement for Patterson. The reaction of the "whites" was more negative. According to him, half of them did not believe in the existence of the hairy giant. However, those who did, such as Tom Slick, a Texas millionaire, set a powerful example.

Tom Slick, a captain of industry, CEO of a number of companies,[1] had a bright and open personality, preferring the company of scientists and researchers to trendy social types. Slick joined forces with Kirk Johnson, another rich Texan, to finance an expedition to the Himalayas in 1957 to search for the yeti: the Slick–Johnson Himalayan Yeti Expedition.

That venture was under the leadership of Gerald Russell, a well-known naturalist. Peter Byrne and his brother Bryan, both familiar with the region, served as guides. The team also included a photogra-

Tom Slick (top) and Peter Byrne. The two were great friends, and have become legendary figures in the search for the yeti and bigfoot.

---

1. Among which the Slick Oil Company, in Houston; Slick Airways in Dallas; the Bailey-Selburn Oil and Gas Company in Calgary (Canada); the Essar Ranch in Texas: a center for the improvement of bovine husbandry, etc...

pher and a film-maker. A former officer of the Nepalese army ensured liaison with the local authorities.

Three months into the expedition, Peter Byrne described an encounter between a simian creature and a Sherpa tracker named Da Tempa, accompanied by a local villager. One night, after exploring the banks of the Choyang River, the two men were walking back to Gerald Russell's camp at night when they heard rustling in the leaves. In the beam of their flashlight they saw the silhouette of a five-foot-tall (1.5-m) creature—hunchbacked, with a pointed head. Its body was covered with a thick reddish coat. It walked straight towards the two men to within some 33 feet (10 meters). Terrified, the men ran away as fast as they could.

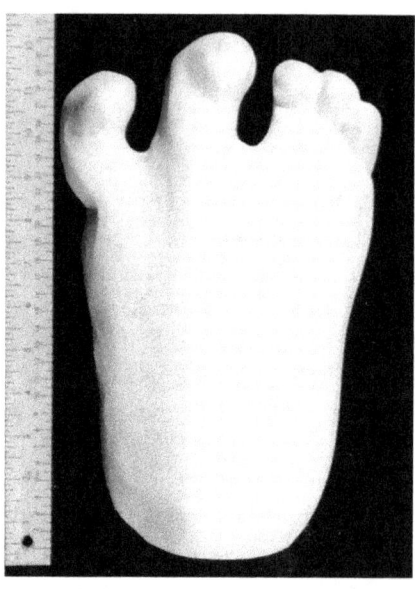

Cast of a possible yeti's footprint (one in a series) found by the Eric Shipton and Michael Ward Himalayan Expedition in 1951. The print was in snow, so the original cast (of which this is a copy) was created using a photograph.

The following morning, Gerald Russell discovered four footprints and took photographs of them. Did they belong to the smaller variety of snowman, the meti, or to a juvenile of the giant yeti species which reaches seven to eight feet (2.0–2.5 m) in height when fully grown? Peter Byrne concluded in his report that the tracks were certainly much smaller than the 10-inch (25-cm) tracks left by an animal that had previously twice visited his camp at night in the Barun Valley.

After four months into the expedition, only Peter and Bryan remained; everyone else had been called away by other business. The brothers carried on until the end of the year (1957) and hiked more than 1,000 miles (1,600 km), returning to Katmandu at Christmas, to rest and resupply, and to wait for instructions from Tom Slick about a continuation of the search through 1958.

Over the remainder of 1957, the sponsors, Slick and Johnson, received monthly reports on the expedition's progress, although nothing of significance was observed or found. At that time, the only connection between Nepal and Texas was through the mail. There were no communication satellites; the Global Positioning System was still a science-fiction idea. Under these conditions and given the poor results so far, was it worthwhile to continue the search?

While Slick was reflecting on this question, Peter and Bryan Byrne were enjoying the somewhat austere comforts of Katmandu's Royal Hotel, the only hotel in the city at that time. Slick's instructions to continue the search came a few days after Christmas. They posed a tough assignment for the Byrne brothers, especially as they included orders to stay in the upper Himalayas for at least another year, or until they could produce the evidence he wanted, which included photographs of one of the creatures. As Peter Byrne later stated:

> in those days we were very keen young men, dedicated to a task to which we were prepared to devote our lives; to what Tom Slick used to call the Ultimate Quest.

In all there were five separate expeditions. The first one was the reconnaissance, which lasted two months (February and March 1957) and consisted of Peter Byrne and Tom Slick, with six Sherpa guides, including veteran Himalayan guide Gyalzen Norbu, 50 porters and a Nepalese liaison officer, named Tree Ratna.

The second, which lasted two months, (April and May 1957) was the first major search, and included Gerald Russell (of giant panda discovery fame [with Ruth Harkness of New York] in the 1930s); a photographer, George Holton, of New York; a movie photographer, German-born Norman Dyrhenfarth, of Los Angeles; and a Nepalese liaison officer, one Captain Pushkar Shum Sher Jung Bahadur Rana, from Katmandu.

The third, June through December 1957, consisted of the Byrne brothers and, through June only, George Holton of New York.

The fourth, January 1958 through December 1958, consisted of the Byrne brothers with six Sherpas from Darjeeling.

The fifth, which began after a Christmas break in December

1958, ran for another 12 months and consisted of the Byrne brothers and their Darjeeling Sherpa guides and, for six months of that year, explorer–photographer Shirley Lawrence of Sydney, Australia. It concluded in December 1959.

During all this time, choosing mobility over comfort, the Byrne brothers operated without tents and nonessential supplies. They lived like mountain shepherds—sleeping in huts or in caves at high altitude, above the tree line. This was certainly a thrilling trek for a pair of adventurous young men eager to prove the theory that there are still some mysterious primates living in the valleys of northeastern Nepal.

> **ANALYSIS OF THE ALLEGED YETI SKELETAL HAND**
>
> **Dr. Desmond Doig:** "It is possible some of the bones are not human, but almost certainly the best part of the hand is."
>
> **Sir Edmund Hillary:** "This is essentially a human hand, strung together with wire, with the possible inclusion of several animal bones."
>
> **Dr. Marlin Perkins:** "This turned ou to be human."

What did the team find? Possible yeti footprints were seen and photographed and three alleged yeti-related "relics" were discovered, a skeletal hand in the Buddhist temple of Pangboche, a single scalp in the great monastery of Thyangboche and another scalp in the Sherpa village of Namche Bazaar, in the Sola Khumbu district. They also found a mummified paw of a snow leopard (an important find at that time).[2]

Realizing the importance of the skeletal hand, but unable to borrow the relic, Peter surreptitiously removed one bone and replaced it with a corresponding human bone. Analysis of the removed bone was inconclusive. Much later, both the skeletal hand and one of the scalps were borrowed for scientific analysis. The results on the hand were essentially inconclusive (summaries are provided above right). The scalp was said to have been made from the skin of a serow (a type of goat), although there was one anomaly—the remains of lice found in the hair were not of the type found

---

2. In those days, the rare snow leopard was still poorly known. It is only after the expedition of George Schaller and Peter Matthiesen that the behavior of the animal became better known. See P. Matthiesen's *The Snow Leopard* and G. Schaller's *The Stones of Silence*.

Some members of the Slick's Pacific Northwest Expedition. From left to right: Ed Patrick, Tom Slick, René Dahinden, Kirk Johnson Jr., Bob Titmus, and Jeri Walsh (Slick's secretary). The photo was taken by John Green.

on this species of goat. It needs to be stressed that only one of the scalps was analyzed, and it has been reasoned that this particular scalp was fabricated for religious purposes (the yeti is considered "sacred" to the people in this region). If they needed additional scalps for their ceremonies, then fabricating one would solve the problem. Perhaps the other scalp was genuine—both were very old—and it is unlikely the temple monks knew the difference between the actual scalp and the fabricated one. We might also note that at that time (late 1950s and 1960s), DNA analysis was not available (not discovered yet), so scientific results were much less accurate.

John Napier was rather severe in judging these discoveries: "No single item contributed one jot or tittle of proof to the Himalayan Bigfoot legend." Nevertheless, Tom Slick did not give up hope. He now turned his attention to the Pacific Northwest regions of North America, where news of a yeti-like creature, bigfoot, had made

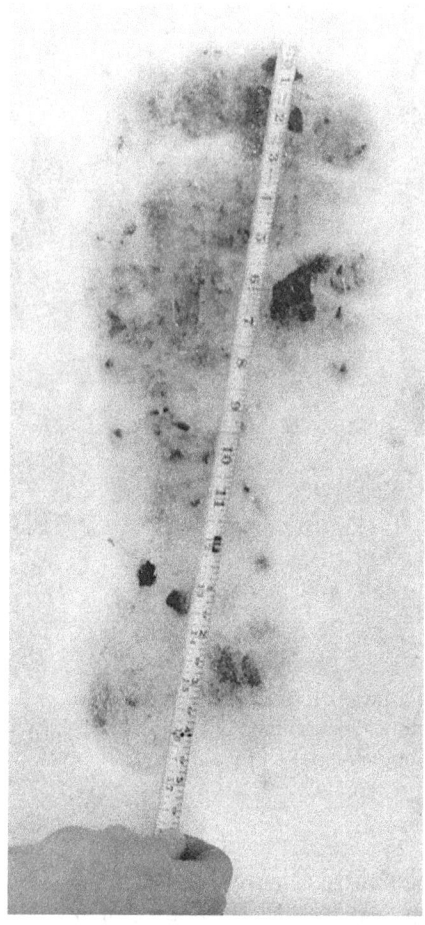
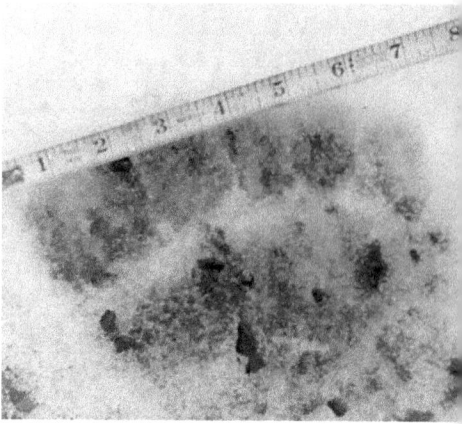

A 17-inch (3.2-cm) footprint in snow found on Offield Mountain (near Orleans), California, in March 1960 by a PNE member. All of the photos taken and other evidence collected by the expedition team were supposed to have been sent to Tom Slick's foundation in Texas. However, it appears at least one member took his own photographs or had extra prints made.

headlines. In the fall of 1959, he organized the Pacific Northwest Expedition (PNE), which included among others, John Green, René Dahinden and Bob Titmus, all of whom were ardent bigfoot researchers.

At the end of 1959, Slick called upon Peter Byrne to oversee the PNE operations. He thereupon left Katmandu, Nepal, and three weeks later, settled into a small motel at Willow Creek, Humboldt County, California. In his own words:

> There, a stranger in a strange land, I set about trying to organize the first Bigfoot expedition, a systematic and sensible approach to the problems of finding out, A) if the things really existed, and B) if so, how to go about finding one.

Thus began Peter's long involvement in a search that is still underway.

The PNE members were certainly well qualified for the task at hand. However, the hunt for bigfoot soon attracted both the wise and the foolish. It is not my place to pass judgment—at least for now—on some "investigators" or their motives. I merely wish to present some unusual people (crackpots, crooks, jokesters) as they were, with all of their idiosyncrasies, as they venture through the grand scenery of the Pacific Northwest—a fascinating aspect of the "human comedy" in the search for bigfoot.

In this connection, I cannot resist the temptation of recalling one of Peter Byrne's anecdotes, typical of those tall tales dear to many American authors, like Mark Twain. Stories such as this illustrate the nature of some of the colorful characters who became "associated" with the Pacific Northwest Expedition.

There were three somewhat mysterious figures who were written up in my files as "associates to the expedition." Inquiries showed that they were engaged in the logging and timber trades and that their interest seemed to be mainly in the area of supplying evidence which they themselves had unearthed in their own searches. One of them lived in Willow Creek at this time. The others lived in Weed, Oregon and Drain, Oregon. The Willow Creek individual was the representative of the trio and in due course I made contact with him. As the ensuing farce was something that I can never forget, I feel that it deserves telling in some detail.

I found the leader of the trio to be a tall, angular man with a hard, thin face, a large bony nose, close-cropped, graying hair and long yellow teeth. Perhaps because of his work amid the constant roar of high-powered logging machinery, he had acquired the habit of shouting. Thus he never spoke, but shouted all the time. Yes! He shouted, ushering me into the house, I had come at exactly the right time. His companions, he roared, his "range and track teams," as he called them, were real woodsmen and at this moment were hot on the trail of a goddam Bigfoot. They expected to capture the bugger at any moment. They were first-class trackers and top woodsmen and they were doing something that none of these yahoos who worked for Slick were able to do, "Shit, man, these guys could track a squirrel up a tree and down again." In a few days he would be calling me and I could rest assured that he would have something to show me then. He shouted me to the door and I left with my ears ringing.

Sure enough, three days later, he called. The phone crackled in my ear. "We have it," he roared. "My range and track team has done it! We've got the bugger! It's only a young one, but it's a real goddam Bigfoot. Why shit, man, it's got hair all over it and the biggest goddam feet you ever did see. You wait until them sciences [scientists] see this bugger, man." When I could get a word in, I asked him when I could see the beast. "Why man," he replied, "as soon as them sponsors of yours come up with our price, that's when." I gently asked him what the price might be? One million dollars, in cash, and they would hand the beastie over, complete with carrying cage.

I called Slick in San Antonio and told him about the situation. He told me to offer them five thousand and to tell them that we would pay it on sight. At first my bellowing friend would not hear it. Then he said that he would confer with his partners. He would call me in a couple of days. Two days later he called and said that they would accept the five thousand but that they would have to have it in advance. Nothing doing, I told him and so we went into two weeks of phone calls and argument back and forth, while we bargained for a look at the creature and they bargain for cash on the nail, sight unseen. Then came an urgent call from Shouter. They were getting into difficulties. The only thing that the young Bigfoot would eat was Kellogg's Frosted Flakes and it ate them by the hundred-pound bag. It was, in fact, running them dry financially, and I would have to do something quickly. There was only so much money available for Frosted Flakes.

I called Slick again and told him about the Frosted Flake problem. He told me to offer them five hundred dollars for a look at the creature and to take a camera along. I did this; they said no, and asked to meet with Slick. He flew up from Texas and we met with Shouter at the cafe opposite the motel at Willow Creek. Shouter described the creature and told us what it ate and so forth. Slick thereupon wrote a check in the amount of $25,000 and put it on the table. When Shouter reached for the check, Slick put his hand on it and slowly drew it back saying, "If you don't mind, we would like to have a look at the thing first." Shouter said, "Okay, I'll meet you here tomorrow at about 9:00 am and then I'll take you to see it." The next day Tom and I went to the cafe as arranged and Shouter showed up with a long, drawn, sad face. Then, as I rather expected, he said the thing had become very ill. It had weakened and was very near death. Rather than let it die, in the goodness of their hearts, they decided to let it go. "You should have seen that bugger run when he got out of the cage," Shouter told us. "But I thought you said he was sick," I responded. Whereupon the Shouter took umbrage at my suspecting his honesty in the matter and did not call me again.[3]

Soon after this burlesque episode, candidly revealing of the nature of some of the locals, Peter Byrne managed to gather two other PNE members—his brother Bryan, and Steve Matthes, a professional hunter from California. From time to time, Tom Slick himself (whenever his business activities would allow) joined up with Byrne's team.

During the next two and one-half years (1960–1962) the PNE found more than a dozen sets of probable bigfoot footprints. Even the most skeptical citizens of Willow Creek and Salyer, California, offered to help, without any resentment towards the "foreigners," (the Byrne brothers being originally Irish).

Unfortunately, the first major bigfoot expedition ended abruptly in the fall of 1962—Tom Slick and his pilot were killed when their plane crashed in a violent storm over Dillon, Montana. Slick's family and associates did not share Tom's interest in the "unknown," so did not wish to further support the PNE initiative.

---

3. Peter Byrne, *The Search for Big Foot: Monster, Myth or Man?* pp. 105–106 and personal communications, July 2008.

Peter Byrne's remarkable book, published in 1975 (Acropolis Books Ltd., Washington, D.C., edition) and 1976 (Pocket Books, New York, edition) is now one of the great classics in the annals of sasquatch/bigfoot studies.

Throughout the PNE years of operation, all findings and reports were sent to Slick's "Southwest Foundation" in San Antonio, Texas. A full "official" report on the expedition was never published.

Peter Byrne thereupon returned to Nepal and took up the profession he had pursued before working for Tom Slick— that of a big game hunting guide. This he followed until 1968, and then hung up his rifle for good and pursued his life-long interest in animal conservation—he helped to found the International Wildlife Conservation Society, Inc., (Washington, D.C.) and later established an animal reserve in Nepal.

In 1970 he again pursued bigfoot research, directing The Bigfoot Research Projects (three separately funded projects that continued until 1996). Peter presently spends six months of every year in bigfoot research, working from a base in Pacific City, on the Oregon coast, and the remaining six months working in wildlife conservation projects in Nepal.

Chapter 6

The Patterson/Gimlin Film

At the time of the Pacific Northwest Expedition, Roger Patterson pursued his own enquiries and closely followed the work of others, including that of Tom Slick's team. He was accumulating bigfoot-related documents—new and archived newspaper articles and special studies. We recall that in 1966, he had published a book on bigfoot, or what he called the Abominable Snowmen of America. He also created the Abominable Snowmen of America Club, and distributed a quarterly newsletter to its members.

Patterson frequently traveled into forested regions on horseback, with a packhorse loaded with supplies. At night, he set up his camp, in the manner of a Hollywood cowboy on the range. He wore a wide-brimmed hat as well as chaps, which protected his legs against thorns and spiny bushes.

He cooked his meals in a metal can, suspended from a tripod over a wood fire. Sitting on a stump, his gun and a canteen handy, he poked the fire with a stick, his sleeping bag rolled up at his feet. Behind him, tied up to a Douglas fir, the horses seemed to be standing guard over him.

The scene is a familiar one—an overdone cinematic cliché. Perhaps the life of a trapper, hunter or gold seeker may appear as a series of clichés—idyllic scenes set in the magnificent scenery of lakes, mountains and forests. All is tranquil, until something happens to break the spell— the apparent peace is only temporary and is often abruptly interrupted. Patterson knew this well, as do all who are familiar with the forest—which reminds me of a story told by my friend Dick Barbre.

In his youth, Dick Barbre was a cowboy for many years in Montana. At the age of 17 he found a job as a logger and set off on his horse to join the logging camp in the hills. One night, after work, he decided to climb up a mountain. It was summer and he wanted to take advantage of the long daylight hours. The undergrowth became thicker and his horse was struggling up the slopes. The hours were passing and Dick soon realized that he was lost, unable to find his way back to the camp.

A superior drawing of a lynx (bobcat) by my good friend Dick Barbre. He became well acquainted with the ways of nature and often expressed his love of the outdoors in artwork.

Suddenly the sky darkened. Black clouds and darkness fell over the landscape. Thunder growled. Dick tied his horse to a tree, fearing that it might run away in a panic. Torrential rain poured on the forest. Wrapped in his sleeping bag, grasping his gun, shivering with cold and fear, and constantly startled by the sharp noises shaking the mountain and the forest, Dick spent an agonizing night.

At the crack of dawn, carefully leading his horse, he started down the hill. Eventually, he heard the voices of his coworkers. Hair in a mess and shivering with cold, he walked towards the foreman, sure that a search for him had been initiated the night before and would be about to resume. But the foreman said, "So, where have you been? It's about time you got to work."

The story is revealing. As always, the storm is unforeseeable, and a fall may happen. Loneliness, intensified by the frightening sounds together with cold, hunger, fatigue, and the possibility of an encounter with wild animals, leads to fear and maybe panic. The attitude of the foreman, apparently lacking all sympathy for the plight of the young man in what he had experienced, was explained a few years later when he told Dick, "That morning, I took no pity on you to make sure you didn't pity yourself. You had to snap out of it! But I knew what you'd been through."

Roger Patterson had probably also lived through that kind of experience in his years as a rodeo rider and bigfoot researcher. Listening to Dick Barbre, I could readily imagine hearing many exciting stories from Roger. However, none would top the events of that day, October 20, 1967, when he was riding with his partner, Bob Gimlin, in the valley of Bluff Creek, in Northern California, which I will now relate.

Just beyond a sharp curve they came upon a clearing, created by a dry part of the streambed. At that moment, their horses reacted in apparent fear; Patterson's horse reared, bringing him to the ground. To the men's astonishment, a massive man-like creature was standing on the opposite side of a shallow creek (Bluff Creek) fewer than 100 feet (30 meters) ahead.

Patterson grabbed his camera and started filming while running toward the creature, moving from left to right. It stood about seven feet (2.3 m) tall, and weighed over 500 pounds (over 226 kg). Its pendulous breasts (it was obviously female) were clearly visible. Dark brown-red hair covered its whole body except its face, palms, and soles of its feet. The creature swung its arms widely and walked with a regular rhythm along the gravel sand bar. What appeared to be a bony ridge stood out at the back of its head; the neck was short and thick; the muscles of its back and shoulders were impressive. The shape of its arms and legs was unusual, as if weighted by the massive muscles seen rolling under its skin.

Patterson kept running, frantically trying to focus his camera on the retreating figure. The creature, first seen briefly face to face, was now seen from the back as it headed toward the forest fringe; then in profile, and then half turned around by rotating its head and

©2007 Michael Rugg

This exceptional artwork by Michael Rugg called *The Moment*, depicts the moment Roger Patterson and Bob Gimlin spotted the creature. Mike sought the assistance of Bob Gimlin, Chris Murphy, and other researchers in creating the scene. It is believed to be very accurate.

This scale model shows where the main action in the film took place. Patterson was standing in the position indicated with a pin at the lower edge of the model. The particular moment depicted is when the creature turned and looked at the two men.

torso—(both turned in unison, like someone who had a stiff neck and back would turn). It gave Patterson and Gimlin a brief, almost casual, yet concerned look. The creature simply kept moving straight ahead, and a back view was again seen as it proceeded into the forest directly ahead and disappeared.

Meanwhile, Bob Gimlin had released the pack horse in order to control the horse he was riding. He crossed the creek, dismounted, crouched down, and with his rifle in hand observed the whole scene. After the creature disappeared into the forest, he proceeded to follow it on horseback, but Patterson, who did not have his horse or his rifle, called him back. The two men then fetched Patterson's horse and the pack horse and both followed the creature's tracks, which led up the creek (opposite of the flow). They found scuffed prints that indicated the creature ran after it was out of their view. They also found a wet half-footprint on a rock. Eventually the tracks led up into the mountains where they could not longer follow them. They thereupon returned to the film site and took movie footage of the creature's tracks. They then went to their campsite—fewer than two miles (3.2 km) away—to obtain

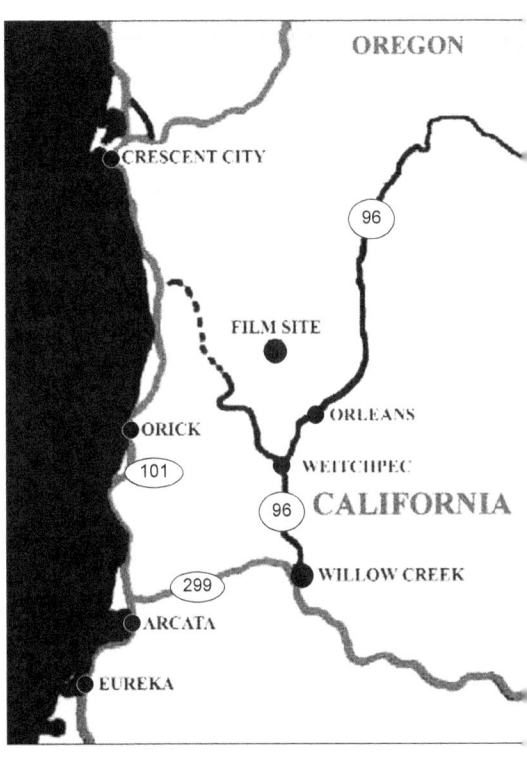

NORTHERN CALIFORNIA

Roger Patterson (left) in 1973 and Bob Gimlin in 2003.

An actual film frame showing the creature when it turned to its right and stared at Patterson and Gimlin.

plaster, and subsequently made plaster casts of a left and right footprint left by the creature.

Highly excited with the prospect of filming a bigfoot (there was no guarantee that the film would turn out) they tethered their horses at their campsite and went immediately in their truck to ship the film for processing. Their plan was to wait for word that the film was valid, and if not, they would stay longer and try again to film a bigfoot.

Upon return to their campsite late that night it started to rain. As the night progressed the rain increased and forced the two men to leave the area early the next morning for fear of landslides. They had great difficulties getting back to the main road as they had to take a different route out—the rain had swollen Bluff Creek and they could no longer cross it to get to their original route.

At the time the two men went to ship the film, they met with a friend, Willow Creek store owner Al Hodgson, and his family. Patterson excitedly related the incident to all and requested Hodgson to contact a British Columbia scientist, Don Abbott, to come to the site. We will discuss the contact aspect in the next chapter.

The Patterson/Gimlin film segment that shows the creature is about 26 feet (7.9 m) long. It is comprised of 953 film frames, 16 mm wide, in color. Patterson had used the first 74 feet (22.6 m) of film on a 100-foot (30.5-m) roll to film scenery and general shots of himself and Gimlin. When the creature was spotted, he had only 26 feet (7.9 m) of film left, and actually ran out of film as the creature disappeared into the forest.

At the beginning of the creature segment, there are a few very clear shots of the creature from behind (i.e., we see its back). Then, when Patterson is running to catch up with the creature, the film images are very poor and unsteady—little can be seen (he actually trips at one point). When he stops running and properly focuses the camera, the creature is revealed with reasonably impressive clarity. When the film frames are isolated (seen as a still image) the massive appearance of a female sasquatch is very evident. One can see clearly its thick body (without any narrowing at the waist), short hair, and a conical skull set on its shoulders without evident appearance of a neck. Its face is somewhat obscure, but certainly has less hair than other parts of its body in general. From what can be seen of its hands, we can reasonably conclude that its palms are free of hair, as are the soles of its feet—there are good shots of the sole of one foot. The legs appear to be thick and muscular; the buttocks very pronounced.

Patterson and Gimlin, two intrepid adventurers, had done what no one before them had been able to do—obtain reasonably clear images of the elusive North American bigfoot.

An Extraordinary Document

One can readily imagine the interest aroused by the Patterson/Gimlin film among specialists as well as members of the general public. It was shown everywhere in North America and it continues today to be analyzed and dissected at zoological and cryptozoological meetings.

Let's hear what Dr. John Napier had to say about it. At the time,

Napier was on the staff of the Smithsonian Institution; his scientific expertise had earned him a solid and well-deserved reputation.[1] While keeping an open mind, he strove for rigor and prudence in his critique. His analysis of the film took place on December 2, 1967, only a few months after the Bluff Creek encounter.

Napier's observations may be summarized as follows:

1. Generally speaking, the gait resembles that of modern man, *Homo sapiens*.

2. The rhythm of the step, the fluidity of the body's movement and the swinging of the arms are greatly exaggerated.

3. In spite of the presence of the pendulous breasts, the gait is more like that of a male human being.

4. The conical shape of the top of the head is certainly not a human shape; it is often seen in male gorillas and male orangutans. The role of this bony crest is to provide an anchor for the attachment of the powerful muscles which move the massive jaws. This is essentially a male feature, rarely observed in females.

5. The anatomy of the creature and the weight that it implies, especially at the level of the chest, suggest that the center of gravity should be higher than that of a man, thus affecting its gait. However the motion indicates that the center of gravity is at the same level as that of modern man.

6. The prominent buttocks, typical of humans, clash with the simian nature of the upper body. The upper part of the body resembles that of an ape, the lower that of a human. It is unlikely that such hybrid creatures could exist in the wild. Half of this creature is certainly inauthentic; in light of its gait, it can only be the upper part.

---

1. British-born Napier was a primatologist specializing in prehuman fossils. He was then director of the Primate Biology Program at the Smithsonian Institution, Washington, DC.

In John Napier's opinion, there is something unnatural about the gait. Why then would a meticulous hoaxer have thus spoiled the effect? Napier supposes that a costumed person, intent on following a straight line, might have been trying to leave clear footprints in the sandy ground while taking giant footsteps. Hence the exaggerated aspect of the gait.

Napier did not reach his conclusions a priori, but only after meticulous and critical examination of the film. This is worth noting. The scientific community's dismissive and disdainful attitude towards unexpected discoveries is more common than one might believe.

A classic example of this attitude is the identification of the coelacanth, the "fish with legs" belonging to a group thought to be extinct for 65 million years, captured alive off the coast of South Africa in 1938. Professor J.L.B. Smith described in his book *Old Four Legs: The Story of the Coelacanth* (Longmans, Green, London 1956)[2] the difficulties he had convincing his colleagues that they were in the presence of a living fossil, in spite of the fact that they actually had a specimen to examine.

---

2. There is also an American edition: *The Search Beneath the Sea: The Story of the Coelacanth*, (Henry Holt, New York, 1960)

Chapter 7

René Dahinden and Peter Byrne: The View from the Forest Floor

We shall now look at to the responses, experiences, and actions of René Dahinden and Peter Byrne regarding the Patterson/Gimlin film—generally treated with scorn by the scientific community in North America in 1967–68.

René Dahinden (1930–2001)

Dahinden was so appalled with North American scientists that he undertook to find professionals elsewhere in the world who might take the film seriously. In 1971 he traveled to Europe (England, Sweden, Switzerland, and Moscow, Russia) to show the film, photographs of footprints, and footprint casts to European scientists. Although he received some encouraging (but qualified) responses, the general outcome was not as he had hoped.

Dahinden was a highly colorful individual, and I believe it is important to know something about the man. The following is an account that was prepared with, and approved by, Wanja Twan, René's estranged wife.

Born illegitimately in Lucerne, Switzerland on August 23, 1930, René Dahinden was placed in a Catholic orphanage at the age of one month. About one year later, he was adopted by a middle-aged couple who ran a stationery wholesale business. The legal requirements, however, for the adoption were never completed.

René in 1993.

Nevertheless, his new parents were fairly well situated, so René's infancy and early childhood got off to a good start. René enjoyed skiing, so he was taken on ski trips, and had the luxury of spending his vacations at a summer home. At about age nine, his foster mother died. His foster father took a new, younger wife within a year or so who, unfortunately, did not take a liking to the boy. So, at age 11, René was sent to a boarding school. In René's own words, "Neither of them wanted me, so I was put in a boys' institution in Lucerne."

Life at the school was fairly good. He attended regular school, worked on the school farm, and occasionally was sent out to work for outside farmers. Nevertheless, the last two years had hardened him to the ways of the world, which was reflected in his character. Remarkably, a little over a year after René entered the school, his natural mother showed up and claimed him. By this time she had remarried and had two children. René joined the family, but his home-coming was a catastrophe. He lasted about four months and was then fostered out to a farming family.

While the recent years had been trying for René, past hardships paled to those he would now face. "Life was hard," he once exclaimed, "There was absolutely no time allowed for play. As soon as I got home from school, I had to start the chores, and I worked at them until bedtime. It wasn't that these people were cruel. They just had no time for affection." He remained at the farm for three years and then tried again to live with his mother and her family. Now 15 years old, he stayed about two weeks and then struck out on his own.

Finding work when and wherever he could, he survived the next three years with little trouble. When he turned 18, he had his mother sign the necessary forms for a passport. For the next five years, he wandered all over Europe, work-ing long enough at one job to get enough money to move on to another one. In Sweden he met his wife-to-be, Wanja Twan, in September 1952.

The following year, René decided to immigrate to Canada, and shipped out in October 1953. He went to work on the farm of William Willick near Calgary, Alberta. While there, he and Willick heard a CBC radio program about a *Daily Mail* expedition to find the yeti. This aroused René's interest and he remarked to Willick, "Now wouldn't that be something—to be on the hunt for that thing?" Willick responded, "Hell, you don't have to go that far; they got them things in British Columbia."

René pressed his boss for more information and found out that he was not kidding. René moved to Williams Lake, British Columbia the following spring, where he found work at a sawmill. He spent his spare time doing sasquatch research. Wanja came to Canada in 1955 and the two married the following year.

Their first son, Erik, was born later that year. In 1958, René operated a boat rental service on Harrison Lake, B.C. Wanja worked for a local bank. In those years, skeet shooting was held on the muddy fringe of Harrison Lake by a traveling skeet club. Wanja commented on the lead pollution caused by the spent shot, stating that the shot should be retrieved. Her comment gave René an idea, and he thereupon recovered and sold several tons of it, making a reasonable profit. The couple's second son, Martin, was born in 1963.

By this time, René was totally consumed with the search for the sasquatch. Indeed, over the last ten years he had made a name for himself in the field—very little happened in the West "sasquatch-wise" without René's involvement. He left his operation at Harrison and concentrated on lead shot salvaging at gun clubs—eventually working full time, when he was on site, at the Vancouver Gun Club in Richmond, B.C. The club provided him with living facilities on club property.

By 1967, René was spending very little time with his family. He had made a conscious decision that nothing else mattered except the sasquatch, including his family life. Erik was then eleven and Martin four. Fortunately, Wanja was a responsible mother and able to carry on reasonably well without René. The couple formalized the situation with a divorce in August 1967. Although Wanja was not happy with the state of things, she understood that René's passion would give him no peace until he solved the sasquatch mystery. The two remained good friends to the end.

René's catamaran, now high and dry on his son Erik's farm in Enderby, B.C. The crest on the front (inset) was very appropriate.

Certainly a highly dedicated researcher, René spent as much time as he could at public libraries and generally reading everything he could find related to the sasquatch, and indeed other crypto creatures. He even went in search of Ogopogo, the lake monster said to live in Lake Okanagan. However, as he was amassing documentation on the sasquatch, people he consulted tried to dissuade him, "Come on! It's an Indian legend, nothing more!"

René readily admitted that at that time, impetuous young adventurer that he was, "I was not in a position to evaluate what I heard or read." Nevertheless, as the years went by, he examined reports, interviewed witnesses, and verified stories provided to him. His continuing research further enhanced his curiosity. In 1956, he joined forces with John Green, whose initial publications reflect the discoveries made jointly by the two of them over a period of thirteen years.

Dahinden participated briefly in Tom Slick's Pacific Northwest Expedition, and was paid a salary of 350 dollars per month, with accommodation and food provided. However, he suspected that some of the members of the expedition were exploiting Slick, who spent fairly liberally on sasquatch research. Their greed disgusted René; he also felt that the team lacked organization. He soon left the PNE to continue research on his own, eventually working with a variety of different partners.

What was Dahinden's part in the wake of the bigfoot encounter on October 20, 1967 at Bluff Creek? The story has been told and retold so many times that numerous errors have crept in. This is not to say that the version now presented is error free, but it is believed to be the correct story.

On his way to ship the film, Patterson arranged with Al Hodgson to have him contact Don Abbott, a scientist with the British Columbia Provincial Museum, and request him to come to the site with tracking dogs. Abbott had examined tracks firsthand in late August 1967 on Blue Creek Mountain, California (near the film site) so was considered the best bet, over asking a scientist to come in "cold." Unfortunately, Abbot did not wish to go to the site. He stated that he would wait and see the film. Nevertheless, he did contact John Green and relay Patterson's request for tracking dogs, but John was unable to accommodate this request (i.e., get dogs and go to the film site with them). Green, however, contacted René Dahinden, who was in San Francisco at the time; René immediately headed for Willow Creek and met with Al Hodgson at his store. Patterson called the store from a town about 26 miles (40 km) away and talked with Dahinden, bringing him up to date on events. Dahinden thereupon headed for Yakima to see the film. He regretted to his dying day that he did not go directly to the film site.

Don Abbott preparing to "lift" one of the prints found on Blue Creek Mountain, California in August 1967.

During this time Green and another researcher, Jim McClarin, were en route to Yakima to see the film, which was to be shown at Al DeAtley's home (Roger's brother-in-law who had received the film in the air shipment and had it developed).

In the eyes of Dahinden and Green, and many others as time went by, there was no doubt that the film was genuine. I have personally observed that sasquatch's gait is smoother than that of *Homo sapiens*. Throughout the gait, the knees remain bent, giving the motion a more fluid appearance. In contrast, humans straighten the leg, and then bend the knee, and so on. On the whole, however, the creature appears to walk much like a human. It was, in my opinion, probably a hominid.

The Patterson/Gimlin film added an extraordinary valuable document to René's already bulging

The note left for René by the hotel desk clerk in San Francisco. René kept everything. When asked by Chris Murphy why he did not use a Bic lighter rather than matches to light his pipe he said, "Because I just can't throw them away when they are empty."

files. It was perhaps definitive proof of the existence of the sasquatch—René certainly thought so.

On December 14, 1969, René was found in the company of Ivan Marx in the sparsely populated northeast corner of Washington State (on the shores of Lake Roosevelt, near Bossburg and the Idaho border). The tracks that they found there became both famous and the source of heated controversy. The footprints, first found in snow on a steep lakeshore, crossed a railway track, a road and a fence, led into the forest, and then returned to the lake. The left footprint, 16.7 inches (42.5 cm) long and 5.3 inches (13.5 cm) wide was shaped normally; the right foot, however, was deformed and showed only four digits clearly (the missing toe is thought to have been crowded up so that it did not register, although a slight impression is seen). After examining the tracks for most of a day, Dahinden came to the conclusion that they belonged to a living being with flexible toes. He counted a total of 1,049 prints.

It is noteworthy that Dr. John Napier believed these prints to be authentic (although he stopped short of stating that the maker was a bigfoot). The deformation of the right foot, be it congenital or acci-

René Dahinden holding the casts he took of the "cripplefoot" prints found near Bossburg, Washington, in 1969.

dental, was such that, in his words, "It is very difficult to conceive of a hoaxer so subtle, so knowledgeable—and so sick—who would deliberately fake a footprint of this nature."

From a biological perspective, the prints were convincing, partly because of the anomaly presented by the right foot, and partly because of the normal appearance of the left foot, in speculated function as well as in structure.

I recognize, along with columnist/author Don Hunter, that, "René Dahinden is [was] the model of a man with a mission,"[1] it is not surprising to find that his convictions became firmer—proof of

---

1. See *Sasquatch,* Don Hunter with René Dahinden (1973), p.78.

the existence of the sasquatch was accumulating—partly as a result of his own investigations and also of those of his colleagues.

Peter Byrne

We will now turn our attention to Peter Byrne who was in the jungle of southwestern Nepal on the day that the Bluff Creek creature was filmed. His personal journal records that over the previous two days he and his client, a retired colonel from Tennessee (a heavy smoker and bourbon drinker), had been tracking a leopard.

On the morning of the third day, after two hours of tracking, the two men heard the howling of the animal, still unsuspecting of their presence, in spite of the colonel's heavy asthmatic breathing. There was no sound and the big dog leopard, his black-rosetted coat of orange magnificent in the morning sun, was frozen in position. The colonel, unable to aim, pointed a shaking finger and shouted: 'That's a leopard!' The great cat disappeared into the forest.

Such was this amusing incident which marked that October day in the Himalayas, while 10,000 miles (16,100 km) away in North America the strange encounter with a sasquatch was taking place. I also wish to take advantage of this short digression to call attention to Peter Byrne's professional skills as a hunting guide, his intimate knowledge of the terrain and tracking techniques, plus his long experience with animals and people, without forgetting his healthy sense of humor.

Back in the USA, Peter Byrne viewed the Patterson film and asked himself the obvious question: Was this a real, living wild creature or a man wearing a costume? After consulting a variety of specialists, he eventually (1973) showed the film to the chief technician of Disney Studios, in Burbank, California, who declared:

> If it is a fake then it is a masterpiece, and as far as we are concerned the only place in the world where a simulation of that quality could be created would be here, at Disney Studios, and this footage was not made here.

During a personal interview, which Peter granted to me in April 1994, he added that the Patterson movie still had to face the gaunt-

Peter Byrne (right) and René Dahinden in 1996. The two met after Byrne had arranged for a highly professional scientific examination of the film by Jeff Glickman, a certified forensic examiner. The final report, *Toward a Resolution of the Bigfoot Phenomenon,* stated that no evidence could be found that the film was fabricated. Indeed, to the contrary, many details were observed that indicated the creature filmed was a natural being.

let of the latest digital analysis techniques. One could not be sure that the film was genuine until this costly process had been completed.[2] However, he did not think that there was trickery—such a complex hoax did not seem possible. To his knowledge, the authors

---

2. Considerable work with digital analysis is reported in Chapter 6 of Chris Murphy's *Meet the Sasquatch,* Hancock House, (2004).

of the film, Patterson and Gimlin, were not capable of producing (fabricating) such an elaborate sasquatch encounter.

Byrne was personally acquainted with Patterson and appreciated his genuine interest in bigfoot. The book Patterson wrote, illustrated, and published was a clear demonstration of his dedication.

Byrne visited Patterson a few weeks before he passed away. Patterson was in great pain—in the throes of Hodgkin's disease; he weighed only 100 pounds (45 kg). "I will never forget one of his last confessions," Peter told me. "Patterson said, 'We had a gun and we should have shot it; then people would have believed us.'"

Right up to his last days, Patterson endured the skepticism often expressed with regard to his film. Widely shown, the film brought him some compensation in the form of admissions and copyright payments. After his death, the film rights passed to his family. His partner, Gimlin, never received the compensation to which he was entitled. Although he was somewhat bitter about that, Gimlin never denied the film's authenticity. Peter Byrne thought that Gimlin could easily have sought revenge by claiming that the film was staged.

Byrne also noted that there were some disadvantages associated with the location where the film was taken (the film site). On a Friday afternoon, there was a chance of meeting hunters or hikers in that particular area. There exists in that region many isolated sites that were much better suited to stage a hoax.

***

I need to now mention that Dr. Grover Krantz[3] interviewed Bob Gimlin soon after Patterson had died. Gimlin assured Krantz that there was absolutely no basis to support the hypothesis of a hoax being perpetrated by Patterson. One can certainly reason that an accomplice in a furry suit working with Patterson alone (i.e., without Gimlin's knowledge) would have been in danger of losing his life. Although Gimlin and Patterson had agreed (made a pact) not to

---

3. Grover Krantz was a professor of anthropology at Washington State University, Pullman, Washington.

Roger Patterson (right) is seen here with René Dahinden. Roger liked to invent useful devices, and the display on the roof of his van shows his "Prop Lock" invention, patterned after the bread bag clip. The device was attached to fruit tree props (poles) and was a great improvement over the standard "Y" props, as it would retain branches in moderate or heavy winds, yet could be easily removed when required.

shoot a bigfoot, in the case of an encounter with the creature (real or fabricated) could Patterson foresee Gimlin's reaction? Would Gimlin, in a state of panic, or perhaps thinking that his companion was in danger, not have used his rifle?

Finally, the question must be asked as to Patterson's technical capabilities to undertake such a fabrication. Both Grover Krantz and Peter Byrne did not think so. When Krantz, commenting on specific points in the film, broached some technical aspect of biomechanics with Patterson, he was immediately left behind, just like a young wide-eyed student caught beyond his depth. In order to create even a marginally convincing fake bigfoot like that seen in the film, Patterson would have needed at least a minimum understanding of biomechanics.

In spite of all the questions and suppositions, mystery still lingers over both the observed being and the observers. In fact, who was Patterson? I have often studied his face on photographs. What I have seen is an aging cowboy, looking somewhat like Montgomery Clift in John Huston's film, *The Misfits*. One of those uncomplicated adventurous Americans—a rodeo rider, a bit of a stunt artist, easily able to adapt to life's circumstances, and happy in his little ranch without much need for luxuries. We see him with his wide-brimmed hat, his sheepskin vest and high-heel boots, standing against the wall of his modest log-and-planking house. I can readily imagine him riding along mountain streams, under the giant trees of the forest, keenly aware of sights and sounds, at ease in the familiar wilderness.

Was he, as I would readily believe it, a loner who never gave up in his quest for the mysteries of the forest? One day the forest sent him a sign in the form of the sasquatch sighting, removing him even further from the fellowship of mankind.

Western/central Washington State. The entire state is about 17% larger than England and Wales combined. Nearly one-half of Washington is forested. The state ranks second (behind California) in the number of reported sasquatch sightings/incidents in the United States. British Columbia ranks first in all of North America. (Image from Google Earth © 2008, TerraMetrics, Tele Atlas, Europa Technologies.)

Chapter 8
Encounters in Neah Bay and Elsewhere

We now journey to the northwest about 250 miles (400 km) as the crow flies, to the extreme tip of the Olympic Peninsula. Endless beaches, strewn with tree trunks, line the wind and wave-swept coast. The cliffs erode under the pounding of the waves, baring the roots of giant conifers, gray ghostly silhouettes soon to fall into the sea.

Looking to the east along Juan de Fuca Strait, we can see container ships and tankers bound for the ports of Seattle or Vancouver. Across the strait, 20 miles (32 km) away, loom the mountains of Vancouver Island, in Canada. Sometimes, in the cold waters, one can glimpse a pod of orcas.

We are now standing at the extreme western tip of the continental United States. But to really experience the Pacific Ocean in all its might, one must cross the Makah Indian reservation, leave the car at the end of the road and walk a few miles to the shore through the forest.

In the early sixties, a young member of the US Air Force, living on the reservation with his Native wife, corresponded with Ivan Sanderson. He related how he had seen footprints on a nearly inaccessible beach guarded by a dense forest. The prints were 18 inches (46 cm) long!

One summer, a fishermen friend of his, also living on the reservation, had been awakened by the splashing of a creature cavorting in the swamp next to his garden. Flashlight in hand, he had glimpsed a very tall creature disappearing into the forest. In the morning, the young airman, gun in hand, had examined the shores of the swamp and discovered large tufts of hair left on the bushes. "I have hunted and killed quite a few bears around here but the hair that I found that day was definitely not from a bear. For one thing, there was a couple of hairs that I measured to be close to 14 inches (36 cm) long and these hunks had a very strong odor unlike any bear that I have killed," he wrote to Sanderson.[1]

---

1. Ivan T. Sanderson, *Abominable Snowmen: Legend Come to Life*, p. 143

At some later date, the young airman was bear hunting, walking one evening along a deserted forest road. He had just walked a dozen miles (19 km) and knew there was no one living within about twenty miles (32 km). He lay down on a stump to rest. Soon, a sharp cry arose, similar to that of a baby, and continued for nearly an hour. After carefully listening to the sound, the hunter decided that it could not possibly be a cougar. The cry ended abruptly and the hunter hastened home. He later returned to the site, without noticing anything new.

In another incident, on October 21, 1972, five men made recordings of unidentified vocalizations in Northern California. Hidden in a shelter of logs and branches, they were resting on their sleeping bags. Moonlight filtered through the branches, reflecting on scattered snow patches. Below, a stream provided a steady background noise. Cameras were at hand, and tape recorders ready to roll—connected to two microphones, one inside the shelter, the other in a tree a hundred feet (30 m) away. For an hour, the outside microphone captured a long vocalization. I listened to the recording, sent to me by Peter Byrne in 1995. There were horrifying screams, sounding at first like monstrous dogs. After listening to the tape a few times, I reasoned that the animals might have been monkeys. I imagined a verbal battle between two baboons, lips curled aggressively, threatening each other without any intention of ever coming to blows—two beasts testing each other, perhaps males disputing a territory, each trying to scare the other by its screams.

The five men estimated that the screamers might have been about 150 feet (50 m) away from the outside microphone. Nevertheless, the sounds seemed very near.

A second sequence included a kind of dialogue. One of the men came out of the shelter and said, "Here! Come over, we won't hurt you!" Something uttered a series of barking groans as a kind of response. The men who made the recording thought that a hoax was always possible, but quite unlikely. Given the distance of the source, it would have taken a large and sophisticated amplification system to achieve the acoustic effects recorded, especially at such high volume; not to mention the need for one or many technicians. Without knowing ahead of time the exact date and itinerary of the researchers' expedition, who would hide in the snowy forest in wait for a party of gullible explorers? Once more, the idea of a practical joke doesn't make sense, given the altitude, the climate, the isolated location and the difficulty of carrying equipment to the site.

Nevertheless, fraud is easy for a skilled technician, especially for one also familiar with the local wildlife. If one is to see everything as a hoax, one must admit that those hills are loaded with impostors.

***

"It was Halloween Night of 1970. I was twenty-three years old, camping near Mount St. Helens. I couldn't sleep, and then the cries began."[2] Halloween night, All Hallows Eve, the night of witches and goblins. Moved by the spirits of the dead, children dress up as witches and vampires. They knock at the door—Trick or Treat! Sitting on the window sill, a carved pumpkin lit by a small candle protects the house from evil spirits.

On that very night in October 1970 Robert Pyle was far removed from children's games. After climbing with his wife the snowy slopes of Mount Loo Wit,[3] they had just slipped into their sleeping bags near the shore of Spirit Lake. As a student of natural history—he was later to obtain a PhD in ecology from Yale University—Robert was familiar with the songs and calls of the area's birds and mammals.

"At first the calls resembled shrill barks, but not canine in the least. Then they accelerated in tempo, grew in volume, and rose in pitch until they sounded more like whistles." Robert reviewed in his head the list of animals likely to be the source of this strange concert: moose, coyote, red fox, lynx, puma, owl. Nothing fit. The extraordinary recital ended with what sounded like the cries of a newborn baby. His wife slept, but Robert stayed awake, listening. Shortly before dawn, the calls ended and he fell asleep.

When he opened his eyes, there was nothing but snow and silence. The echo of the mysterious concert was still ringing in his ears. "Spirit Lake was lustrous in the clear morning," wrote Pyle,

---

2. Robert Pyle, *Where Big Foot Walks*, p. 11.
3. Loo-Wit: one of several variant spellings for a Native American word given to Mount St. Helens. It means "smoking mountain" (Wikipedia).

"and St. Helens an almost clichéd image of mountain glory, its glacial shadows outlined and emphasized by the oblique rays."

Robert read the ancient language of tracks in the snow—the runes of nature—recognizing the footsteps of small animals, jays, rodents, hare. Nowhere in the snow was there a trace of any kind of giant. The idea of a hoaxer hiding at some distance from his camp never crossed Robert's mind.

Once the motors of all-terrain vehicles and snowmobiles have been turned off, all sounds are muffled by the snow, and silence returns to the mountain and the forest. Within such an acoustic shrine of quiet whiteness, an unusual melody would certainly make a deep impression on the auditory memory of one familiar with the sounds of nature.

Thus, the sasquatch's vocalizations echo through the forests and the mountains of the Pacific Northwest, from British Columbia to Oregon. Some have heard them; a few have even recorded them.

In Elma, Grays Harbor County, Washington, I was invited to hear the stories of some eye witnesses—we shall return to them—and I also listened to a tape recording, made by some Indians in Puyallup County.

At the beginning of the tape, one hears some kind of barking, either by a dog or an angry monkey. These rapidly mutate into long ululations from a rounded mouth, with the tongue moving towards the palate to create regular modulations. These screams continued without a break, as if the creature never caught its breath—unless many of them might be taking on the tune, which seemed to emanate from a distant source, spreading over the valley with great force. The listener experiences a feeling of strangeness and unease when imagining the scene—seeing oneself walking along a river or a forest road, suddenly frozen on the spot at hearing these most inhuman screams. At that moment, the visitor to the forest feels vulnerable; he is no longer alone. He faces the presence of another, but what or who?

In spite of many such anecdotes and dozens of pages of notes, there is no easy answer, as one may gather from the great number of newspaper articles, bulletins from cryptozoological societies, books, and television shows. What really matters is the presence suggested by all these reports, selected from incidents occurring between 1960 and 1975.

It's too early to draw conclusions. All we can say is that the events related by the witnesses, be they real or imaginary, had been experienced in some way. This simple introduction to the "sasquatch phenomenon" has allowed the reader to become acquainted with some of the players working to explain the mystery—Peter Byrne, René Dahinden, John Green, Grover Krantz—and also to get a feeling for the vast scale of nature where the action takes place: the Pacific Northwest, from British Columbia to California.

Part II

My Pacific Travels, Insights and Experiences

I have made several trips to the Pacific Coast of North America and have therefore experienced firsthand the vastness and ruggedness of this remarkable region. I also learned of some of the region's history—particularly with regard to its development and its natural resources.

Although what follows in this book part is primarily a diversion from the subject of sasquatch and wild men, I wish to present it for the benefit of those readers who have not experienced North America, or for residents who may be unaware of the knowledge I wish to impart.

Chapter 9

The Pacific Northwest Seen from the Columbia River

Does the noise of a falling tree have any meaning in itself when no one is there to hear it? In the pages to follow, linking nature with the observer, I will describe for you the region as I saw it, with the people who inhabit it.

"What'll be here to keep these people going," a man with baggy overalls and a set of stickery whiskers is saying, "when this dam job is over? Nothing? No, mister, you're wrong as hell. What do you think we're putting this dam in for anyhow? To catch water to irrigate new land and water this desert-looking country here.

"And when a little drop of water hits the ground anywhere out across here—a crop, a bush, sometimes even a tall tree comes jumping out of the dirt. Thousands and thousands of whole families are going to have all the good land they need, and I'm a-going to be on one of them little twenty acres!"

"Water, water," a young man of about 20 or so, wearing a pair of handmade cowboy boots talks up. "You think water's gonna be th' best part? Well, you're just about half right, friend. Did you ever stop to think that th' most, th' best part of it all is th' electric power this dam's gonna turn out? I can just lay here on this old, rotten jungle hill with all of these half-starved people waiting to go to work, and you know, I don't so much see all of this filth and dirt. But I do see— just try to picture in my head, like—what's gonna be here. Th' big factories makin' all kinds of things from fertilizers to bombin' planes. Power lines, steel towers runnin' out acrost these old clumpy hills—most of all, people at work all th' time on little farms, and whole bunches and bunches of people at work in th' big new factories."

"It's th' gifts of th' Lord, that's what'tis." A little nervous man, about half Indian, is pulling up grass stems and talking. "Th' Lord gives you a mind to vision all of this, an' th' power to build it. He takes it away—if we don't use it right."[1]

---

1. Woody Guthrie, *Bound for Glory*, (1943), pp 249 ff.

Roll on, Columbia!

For years, Woody Guthrie's lyrics echoed in my head: Roll on, Columbia. Each time the catchy waltz rhythm sounded in my ears, I would dream of the gigantic river. My three-volume Larousse dictionary said nothing about it; it was Woody who spoke of the river better than could any dictionary.

The world has seven wonders, so the travelers always tell,
Some gardens and some towers, I guess you know them well,
But now the greatest wonder is in Uncle Sam's fair land,
It's the King Columbia River and the Big Grand Coulee Dam.

In 1933, Franklin Delano Roosevelt was elected president. He took on the task of restoring the American economy reeling from the stock market crash of 1929. One of the major projects of the New Deal policy was to develop the hydroelectric power of the northwest rivers to generate cheap electricity.

The voters were faced with a choice between public and private companies to undertake the work. The private firms organized meetings and brought in Hollywood artists to draw in the public. In response, the Bonneville Power Authority[2] called upon itinerant artist Woody Guthrie. Woody's growing success had given him a taste for the amenities of big cities, like Los Angeles and New York, but soft beds and fancy dining left him uncomfortable, especially since his friends and family kept "wandering around over the West like a herd of locoed buffaloes." He had taken his leave of "phoney, bigshot producers," opting for the freedom of the open road.

---

2. The Bonneville Power Administration is a federal agency under the U.S. Department of Energy. It was created in 1937 to market electric power from the Bonneville Dam, the first and lowermost dam on the Columbia. In Woody Guthrie's words:
At Bonneville now there are ships in the locks
The waters have risen and cleared all the rocks,
Shiploads of plenty will steam past the docks,
So roll on, Columbia, roll on.

The Bonneville administrators wanted to retain Woody as a "public relations consultant." They hired one of his friends, Alan Lomax—later to become a well-known folksong historian—to find him. As soon as he heard, Woody traveled to Portland and signed on the dotted line. For a whole month, he was chauffeured around the Columbia basin and drank in its majestic beauty. He studied the proposals and assessed the goals of the government's plan. He admired the irrigated fields and orchards bordering the river with the eyes of an Oklahoma boy used to parched lands under a blazing sun.

With the dust bowl of Oklahoma in mind, Guthrie fantasized about the metamorphosis that would follow the construction of the great dams. He wrote 26 ballads and recorded them in as many days. They were played on the radio and at public meetings. The government plans received massive support.

Woody shared the enthusiasm felt by Lewis and Clark, and by the pioneers of the Oregon Trail. He knew how to translate it into simple and direct prose. He compared the salmon to voters: "Twenty million salmon can't be wrong." Alas! There was no good news for salmon in this whole business. But even today, we can still hear the voice of the mighty river when listening to *Roll on, Columbia.*

I have to confess that I had already fallen under the spell of North America's first great river during my sojourn in the United States in 1965–66. Prepared for it through my reading of Mark Twain, I had expected the encounter with anticipated wonder. The Mississippi did not disappoint me.

When I returned to France to live on the shores of the Loire River, near Nantes, I found another river, of the same nature. I took pleasure in finding similarities, real or imaginary, between the modest Loire and the grand Mississippi.

As the years passed, without ever forgetting my first fascination, Guthrie's melody became the seed of yet another expectation.

Chapter 10

The "Fabulous Beasts" Conference & Introducing "Prospector Ed"

In July 1990, I was invited to participate in a conference on "Fabulous Beasts: Fact and Folklore," held over three days in the bucolic setting of the University of Surrey, in Guildford, UK.

The conference, organized jointly by the British Folklore Society and the International Society of Cryptozoology (the ISC) brought together about 20 speakers on a wide variety of topics. William M.S. Russell, of the University of Reading, UK, spoke on Greek and Roman monsters; Isao S. Uemichi, of Aichi University in Japan, talked about the snakes and dragons of his country; Adrienne Major, of Princeton, New Jersey, expounded on the origins of the griffon; Jan-Ojvind Swahn, of the University of Lund in Sweden, addressed the mystery of the monster of Lake Storsjon; Bernard Heuvelmans—president of the ISC, and inventor in 1950 of the word "cryptozoology"[1]—presented a study of the metamorphosis of unknown animals into fabulous beasts, and that of fabulous

Bernard Heuvelmans with one of his friends.

---

1. Cryptozoology is the study of animals as yet unrecognized by science for lack of definite material evidence. The International Society of Cryptozoology (now disbanded) brought together an eclectic assortment of zoologists, biochemists, linguists, folklorists, mythologists and amateurs interested in mysterious animals.

beasts into known animals. As for me, I spoke on the fabulous beasts of our times.

One presenter who captured the attention of the participants was Ed Fuchs ("Prospector Ed"), who spoke on "The Great Hominid Bipeds Described by the Colville and Spokane Indians." Ed had a commanding presence: of average height, sturdily built, he was shod in leather boots, sported an aloha shirt, and wore a big gold ring on his finger. He spoke in a strong voice, his American accent flavored with colorful rural expressions. In broad strokes, he described the area of his study—Okanogan County in northern Washington State. That's where he lived, in Riverside, a village of 545 inhabitants facing on one side the Okanogan River and the Colville Indian reservation, and on the other Highway 97, leading to the Canadian border, 40 miles (64 km) to the north.

In the course of his enquiry on sasquatches, Ed Fuchs had interviewed the Indians of the neighboring reservation as well as those of another nearby reservation, the Spokane tribe. He described the difficulties he encountered, associated mainly with the age of the interviewees: senior elders, hard of hearing, toothless, bent over, speaking English only poorly. One can understand why Ed, approaching 60 himself, must have had a certain degree of anxiety as to the nature of the information he collected. Had he correctly understood the words of his informants?

It goes without saying that the learned audience, warmed up by the preamble, was eager to hear more. The conference proceedings summarized his presentation as follows:

> He reviewed Sasquatch beliefs in two American Indian groups, based on his own fieldwork of 7 years. The legends and folklore of the Spokane and Colville Indians abound with stories of large bipedal hominids, with whom they reportedly interact and share a common habitat. Mr. Fuchs presented numerous case studies based on his own record of interviews, and concluded that "large bipedal hominids existed, lived among, and interacted with the Indians from ancient days right up to very recent times."[2]

---

2 The ISC Newsletter, Vol. 9, No. 3, Autumn 1990.

In the evening, after his talk, Ed held forth at length. We were sitting in a pub in Guildford with Bernard Heuvelmans, pints of ale on the table. Ed's talkativeness, his loud voice, and his cowboy hat drew the attention of the crowd, whose attitude swung between disdain for the Yankee accent to fascination with the topic of the conversation. Ed spoke eloquently about his homeland, east of the Cascade Mountains, which parallel the Pacific coast. He rattled off the names of rivers and places with Indian-sounding names: Wenatchee, Chelan, Okanogan, Omak, Nespelem, Tonasket, Spokane—so many seeds scattered in the fertile field of my imagination.

He also left me his business card:

Ed Fuchs' business card (with his likeness).

Chapter 11

The Columbia in the Winter

In February 1991, I accompanied a group of computer-science students from the Université de Nantes on a trip to America. The trip included a week in Seattle, visiting local businesses. My Dutch friends, Pieter and Tisch, welcomed me in their home overlooking Lake Washington. One can go by boat from that lake to another, Lake Union, and then through a series of locks to Puget Sound—the Pacific Ocean. Puget Sound, of course, is not the open ocean, but a long and deep fjord leading to the ports of Seattle and Tacoma. A plethora of nearby islands are linked by a fleet of ferries. The city's geography is bewildering; it is difficult to find cohesion in such a collection of lakes, canals, hills, parks, islands and freeways.

The plane landed at Sea-Tac International Airport on a Saturday night. In the morning, bright sunshine reflected from snowy Mount Rainier (14,700 ft [4,400 m]) emerging above the clouds like a print of a Japanese volcano. I breakfasted on a stack of pancakes splashed with maple syrup. Pieter then suggested we go for a ride. We drove south on the western shore of Lake Washington. After a few miles, we drove by the Boeing airfield. I asked Pieter to stop—a shiny red pre-war Boeing Stearman was about to take off. A young man wearing a flight jacket walked by. I asked him if it was possible to go on a sightseeing flight. He took us to the flying-club office where, for a modest price, we booked a one-hour tour.

Sitting comfortably in a Cessna 172 four-seater, we took off towards the east in the direction of the Snoqualmie Falls. Banking sharply at low altitude we enjoyed a good view of Snoqualmie. The pilot then turned back and crossed Lake Washington. We soon flew over the university and then over Lake Union and the seaplane base. Suddenly we were over the sparkling waters of Puget Sound, discovering its many islands, some surprisingly large. A sharp left-hand turn took us along the waterfront, level with Seattle's skyscrapers. A few minutes later, having completed our circuit, we returned to our starting point.

We had familiarized ourselves with the setting, compared distances between various landmarks, and come to understand the magnitude of the Cascade Mountains.

Our week in Seattle was rich in discoveries, be they technological, historic or literary. Names became real people and acquaintances transformed into friendships. Still, I was eagerly waiting for the weekend: Ed Fuchs had invited me to visit him at home.

On that morning, three of us were waiting for Ed to pick us up. Two students had been invited to join us, knowing that Ed's old 4x4 pick-up offered little comfort. Patrick, a timid young man, sat on a crate and a cushion in the narrow space behind the seat. Nugget, the dog, sat on my knees. In the back, under the hard-top canopy, Jean-Yves, cozy in his sleeping bag, his head on a pile of pillows, watched the scenery go by. In the passenger seat, I enjoyed the warmth of the cabin, a blanket (and the dog) on my knees. The double-ignition, six-cylinder motor was alarmingly noisy, especially since there was no muffler.

We were ready for anything and drove north at the speed limit (55 mph) on the I-405 then east on US2, going through a series of small towns—Sultan, Gold Bar and Scenic—before climbing towards Stevens Pass (4,500 ft [1,340 m]). The snowy Cascades seemed to be blocking the way.

Quickly we came out of the plain and entered the forest. The road climbed between two giant walls of Douglas fir, white and green under the snow. From time to time we would come upon a snowplow, blowing a powdery plume off the road.

Some cars had slipped into the ditch, but our studded tires kept us on the road. Occasionally, the motor hiccuped and the car shuddered. The gas pump suffered from intermittent failures. Ed would then push a button triggering a spare pump which he had installed, and the shaking would stop.

After a brief stop at Stevens Pass, Ed drove on. The road aimed southwest towards Wenatchee, after which we would follow the mighty Columbia towards the north as far as Fort Okanogan.

The Great Book

Our travel turned into a colorful lecture, thanks to Ed, who described the scenery. We had followed the meandering beds of the Skyhomish and Wenatchee Rivers. Ed pointed out the gravel bars and the curves of the rivers, the bends carved by the stream. I stared at the rock faces polished by the current, as if hoping to detect a nugget hiding in a crack—we were in the territory of the gold-diggers of the last century. The name of the village of Gold Bar remains a clear sign of their passage.

At Wenatchee we reached the dark waters of the Columbia, crossed by a huge bridge. The river reminds one of the spine of a Great Book, a boundary between two worlds. On the western page, green forests, soaked with rain; on the eastern page, the yellow of sage brush and the black of bare rocks as far as the eye can see.

A geographer might object to my placement of such a boundary. The Pacific slopes of the Cascades are a true rain forest, moist with water dropped by clouds arriving from the ocean; the eastern side, the plateau of the Columbia, is dry, thinly vegetated—a real desert, with rattlesnakes and black widow spiders. The contrast between the two sides is striking, especially in the summer. Once over the pass the temperature rises by 15–25°F (10–15°C). The Cascades rise up to 10,000 feet (3,300 m) near the Canadian border; from north to south, they gradually lose altitude and average about 3,500 feet (1,000 m) by the time they reach the Columbia River valley in Oregon. The mountain chain acts as a screen, capturing the water on its western slope, leaving the eastern side dry. It is the crest of the Cascades that demarks the boundary, not the Columbia River.

The climatic contrast may be strongest at the mountain pass, but a traveler too often consults maps without really studying them; his understanding is based on his sensory perceptions. Awed by the grandeur of the mountain gap—Stevens Pass—the traveler, hemmed in by the tall forest, reacts slowly. Enduring images and sensations saturate his vision even as he travels down the slope. The persistence of images on his retina biases his perception of the scenery. The memory of green at first overwhelms the palette of the new colors which take its place.

It is only when he reaches the Columbia, and not before, that the traveler is struck by the change. He belatedly turns the page of the Great Book. To the serrated partition of the crest of the Cascades, he naturally prefers the deeply incised boundary of the Columbia River.

From then on, the traveler is hypnotized by the grandeur of the scenery. The surface of the river spreads out like a lake, vast and often completely calm. Steep cliffs border the road as it follows the river. Their vertical walls suggest that they were not affected by the erosion that carved the valley. Sometimes a cliff rises above a scree gradually sloping towards the valley. Elsewhere a wooded slope suggests a more stable substrate. Trees readily grow among the broken rocks. Every change in steepness, every topographic detail of the walls of the Columbia Gorge, as the valley is called, takes on some significance.

However, the visitor is usually little concerned with geology, enchanted as he is with the spectacular scenery. Beyond the canyon, bordered by cliffs stepping away like giant staircases, the river spreads out, its gently sloping shores covered with orchards.

Washington State exports apples to the whole world. Interminable trains laden with fruit crawl along the river, like giant caterpillars. Mile after mile, orchards dotted with wind-driven water pumps carpet the shores. The river's water has transformed the near-desert into a blooming garden.

The scenery unfolds: after fields of sagebrush (short bushes, two to three feet [60–90 cm] high, giving the area a Mediterranean appearance), comes the canyon of bare stones and steep cliffs—a scaled-down Grand Canyon—followed by fruit trees and occasional groves of conifers. From time to time, small rocky islands emerge from the river. In a pleasant cove lies a shady park for the pleasure of campers, water skiers and fishermen.

We soon leave to our left the granite crags typical of the Lake Chelan area to reach the site formerly dominated by Fort Okanogan. In a few miles, we will leave the Columbia to follow one of its tributaries, the Okanogan River, northwards towards Riverside.

During the drive, Ed pointed out the pale strata of sediments lying between the dark layers of basalt. These sediments, he explained, contain fossil branches and leaves of sequoias, cedars, maples and elms found only in moist climates; there are also the bones of horses, deer, and predators—the ancestors of cats and dogs. But today, there is basically only sagebrush left.

The reason, of course, is that formerly the Cascade mountains did not exist. In the absence of that obstacle, the rainforests stretched from the Pacific all the way to the Rockies. Or rather flourished during the long intervals between lava flows. Nowhere else in the world is there such a vast field of basalt, through which the Columbia had to erode its bed again and again.

Thanks to the running commentary of my talkative friend, I could now add a fourth dimension—that of time—to my wonder. Gradually, I began to understand not time, perhaps—an unfathomable quantity—but duration, even though it spanned dizzyingly long periods. I could imagine behind their current immobility the great waves that swelled the viscous lava. Its folds and undulations between older basaltic mesas were the witnesses of a spectacle which took place millions of years ago.

Ed explained that before Leonardo da Vinci began to paint people, he tried to understand the human face; first through the study of the skeleton and of the muscles which fashion appearance; then by learning about the tensions, both inner and externally expressed by the face; and finally by studying the changes brought about by the passage of time.

This is also the preferred way to study the Earth. One should first examine its structure—the bedrock and the various rocky formations which define the scenery. Then, the processes of erosion and deposition and all the mechanisms, both slow and fast, which contribute to creating the details of the terrain. Finally, one should consider the evolution of a setting to perceive its former appearance and its future state.

Thus, both Oregon and Washington, now green in the west and yellow-brown in the east were covered with forests before the rise of the Cascade Mountains. The numerous fossils found in the volcanic ashes of eastern Oregon have preserved the remains of forest horses, camels and rhinoceroses.

But before that, lava spread out, in unsurpassed extent; hundreds of fissures spewed out white hot magma, a basaltic eruption without the need of a mountain volcano. From the car, we could see columns of frozen lava along the road leading to the Grand Coulee Dam, to the east. The lava covered 125,000 square miles (323,400 km^2)! The Columbia River had to find its way through and around the wide basaltic plain, dotted here and there by old volcanoes and occasional new ones. Later, the Cascade range rose slowly and intermittently during six million years, at the rate of 12 inches (30 cm) every thousand years, according to geologists, during the Pliocene and Pleistocene epochs.

Suddenly, Ed slowed down and stopped on the shoulder. He turned off the ignition. We stepped out, Nugget first, eager to stretch. Patrick stepped off for a little silence away from Ed's harangue, delivered loudly enough to cover the sound of the motor. Jean-Yves, climbing out of his sleeping bag, walked off his stiffness.

Ed pointed at a long streak in the rocky wall overlooking the road—a vertical crack filled with lava. "There, you see, it's a volcanic dike. It might contain some precious minerals, even diamonds, created by mineralization. One should have a ladder to dig into it. There are fortunes hiding in some such places. Sometimes some guy stops, moved by a sixth sense; he digs and leaves rich. It's happened before."

Thus was Ed sharing his wonderment with me. In front of my very eyes, the scenery was transformed as if a movie animation wizard had recreated in a short time its evolution over the past ten million years. For Ed's trained eye, the earth was hiding another kind of miracle: the discovery of a treasure. I often noticed him looking at a fissure or a crevice, evaluating its potential and taking note of its location in case of a future visit with the right equipment.

"The study of the Earth, the science of geology, is most beneficial for the mind," continued Ed in his didactic mode. " By analogy to Leonardo's method of understanding the human physiognomy, one may look upon this scenery as it if were a portrait. Looking back from its current state, one imagines it in its youth—water roaring through the rocks. Then comes sturdy maturity; the streams have carved their beds, sediments have been accumulated, sand and gravel bars laid out in back eddies. Finally, in old age, erosion has smoothed the hills and rounded off all contours. I suggest that you look at the scenery in the same way as you would greet an old friend."

Ed paused and laughed. "Gee whiz," he said.

"Understanding an old friend is not always so simple. You'll see, some day, when I take you to Grand Coulee. The steep walls of basalt that border the river there look just like the lava and the rivers I have seen in Hawaii. However, that gigantic coulee was carved by the melting of a glacier which spread over north-central Washington State and nearby Canada. That was about 15 million years after the lava flows. There was no epic combat between fire and ice; the lava had by then long been cold and hard. One must be subtle in reading a face or a scenery. One may recognize features, but when it comes to timing events or identifying the origin of rocks, one may readily go astray. Common sense becomes a poor guide.

"For example, it is hard to admit that lava, a very viscous fluid, might cover more than about 100 square miles (259 km^2). But there have been outflows in eastern Washington State that covered thousands of square miles! The mind may refuse to consider that glaciers could have moved boulders the size of a barn. But near Waterville, which we passed earlier today, there are hundreds of such gigantic erratic blocks. The glacier carried them for tens of miles, across the Columbia plateau.

"Common sense recommends prudence and skepticism. Well, here's my advice to you, who aren't a geologist: remember the Red Queen in *Alice in Wonderland*. She told Alice to try, each morning, to believe in six impossible things before breakfast."

It was time to get back on the road. We still had an hour and a half to go before reaching Riverside. It was already dark when Ed drove into the yard of a former primary school, a large two-story building overlooking the road. Over the last ten years, as current owner of Riverside's old school, Ed had been working at repairing broken windows, leaking roofs, missing stairs, burst pipes, frayed wires. I counted two doors and 20 windows on the south-facing side alone.

We entered a large room full of various pieces of equipment: Ed's laboratory, where he analyzes minerals and precious metals— part of his livelihood. Ed took off his hat and emptied the magazine of his Magnum 357 revolver, storing the shells in a separate drawer, a security precaution against mishandling by some careless visitor. He then invited us to his "living room," one of the classrooms of the ground floor, spacious and brightly lit. On the left, there was a desk-library-wardrobe corner; in the right-hand corner, the kitchen; at the back, a wood stove, built out of a hot-water tank; in the center, the owner's bed, and along the wall, a sofa and bunk beds. The rather eclectic furniture included some car seats arranged as a sitting room.

Ed took out of the fridge a bottle of dry and fruity white wine, product of Washington State. Out of the freezer, he handed us large ice-cold glasses which he filled up. "I like it cold," he said.

After we emptied the bottle, Ed looked at his watch. He rushed us out to find a restaurant. After eight-thirty, it is difficult to get served, even on Saturday night. After a few tries, we found a place to eat in Okanogan. The dining room was alive with animated conversations. The guests were a colorful bunch with their hats and western shirts. The dining area was the interior of a barn, the rough wooden walls decorated with a variety of old farm implements. Once dinner was over, Ed opened up a leather purse and pulled out some glass tubes filled with flakes of silver and gold dust. He laid them out on the table, as we watched with fascination and some concern. However, our table was well apart from the others and nobody was paying any attention to us. There, shining in front of our eyes were treasures from the lands which we had crossed. For the finale of this private show, Ed laid on the table a heavy gold ring, bearing a green stone into which had been inserted a massive nugget.

In the morning, when we woke up, it was cold. Logs were quickly stacked into the improvised stove and soon the flames were roaring up the pipe. A quick cup of tea and we were back in the car. We drove to Osoyoos, Canada, just across the border, for a real breakfast—pancakes with maple syrup. On the roadside lay a dead dear, a recent road kill. Knowing that sasquatch are fond of deer, I wondered if they might watch highways for such mishaps and thereby obtain an "easy meal."

We had to return the same day—an interminable drive which lasted until 11:00 p.m, when Ed dropped us off in Seattle. The next day, at six in the morning, we flew back to France.

Chapter 12

The Lesson Continues

A year later, to the day, I woke up shivering. I was facing the wide mountain-fringed valley of the Okanogan. A few small fir trees emerged here and there from the snowbanks. Ed had decided to take me to Spokane, a large city in eastern Washington, near the Idaho border.

Coming out of Omak, a few miles south of Riverside, we drove by a large sawmill and the road (State Highway 155) climbed up. We reached a nearly treeless plateau. Ed slowed down and pointed at some bushes.

"They saw a sasquatch over there, about ten years ago. He ran away towards the woods that you can see back there."

Later, we entered a coniferous forest. In half an hour, we had left the desert to return to the green cover of the tall pines. It became colder as we approached steep-sided Disautel Pass, which reminded me of a coffin. In Nespelem, at the heart of the Colville Indian reservation, we stopped for a few moments in front of the monument to the memory of Chief Joseph.

A few miles later, we joined the magnificent Columbia, now flowing through a deep cleft carved out of dark rocks. Ed explained the scenery:

> This is the work of a giant glacier that came down from Canada, split in two streams by the Cascade Mountains. It was so thick that only the highest peaks emerged from the ice. It is that glacier that carved out the labyrinth of fjords and islands of Puget Sound, in front of Seattle. The ice cap was more than 6,500 feet (2,000 m) thick at the Canadian border, thinning to 1,600–2,000 feet (500–600 m) over Seattle and 330–500 feet (100–150 m) at its southern edge. It also covered Idaho and Montana, smothering all the lakes and valleys. What a gigantic and powerful chisel it turned out to be.

There were four major glaciations over the 1.5 million years of the Pleistocene. Whenever the temperature rose, large freshwater lakes formed behind ice dams; when these melted, the waters rushed downhill like tsunamis, taking whole rockfaces with them. These outflows were only briefly held back by plugs of rocks, mud or sand; they carved their way through all obstacles, leaving deep gouges in the scenery—the spectacular coulees of the central Washington channeled scablands. The Columbia River itself changed its course many times over the years.

Each time the ice retreated, the flora and fauna returned. Incidentally, it is during the second interglacial period that Neanderthals came upon the scene, on the other side of the world. During those long intervals, the climate, the forests and the animals were similar to those of today.

Ed continued:

Let's say that we visit the Pleistocene, but without the glaciers. You can easily imagine them through the impacts of their surges and retreats. The last retreat began only 15,000 years ago. That's only yesterday in geological time! And tomorrow, the cycle may start again.

Don't think, however, that these explanations, authoritative as they may seem, were easily accepted. At one time, they were even considered heretical! It was J. Harlen Bretz, a young geology professor at the University of Washington, who first introduced the term "channeled scablands," and suggested in 1923 that the Columbia Basin had been carved by titanic floods. Many geologists snickered at this outburst of catastrophism. We must recall that at that time some of the public still believed that the Earth was fewer than 10,000 years old, as told in the Bible, and that even though most people accepted that the Earth has been shaped through millions of years, they believed that the processes causing changes had always been slow and gradual. The catastrophic floods described by Bretz were linked by some to the great Biblical flood!

It took until 1952 for Bretz to gather convincing evidence on the ground and to discover a possible source for the immense amount of water required. It then became clear that the Pacific Northwest had not experienced just a single catastrophic flood, but as many as 40, perhaps even a hundred.

The last flood was also the greatest, in keeping with the scale of the last glaciation. An ice dam, 2,000 feet (600 m) thick here, blocked the Columbia's course. An immense lake was created. The day when the dam broke, a 1,000-foot- high (300 m) wave crashed forward at 125 miles (200 km) per hour, carrying icebergs as far as the Oregon coast, nearly 400 miles (600 km) away. In Grand Coulee, even the lava was torn off and carried away with the flood.

Then the glacier receded. The river rid itself of the residual muds and its clean gravels became the spawning grounds of myriads of salmon.

Chapter 13

Grand Coulee and the Great Dam

As we proceeded on our journey, we approached the houses of Electric City (300 inhabitants), a village neighboring Grand Coulee (population 1,500). After a short descent, an immense wall, gray as a stormy sky, came into sight: the dam rose, menacing, stretching between the dark red walls of the canyon. One can hardly imagine the cataclysm that would follow the collapse of this gigantic dike, three times as massive as the great pyramid of Cheops.

Grand Coulee looked like a prosperous little town, with well-kept houses on shady streets lined with tall trees. Who could imagine today the drought that dominated the plateau before the building of the dam? Without irrigation, the weeds and the sagebrush quickly would take over.

Although some early colonists of the nineteenth century thought that the "Interior Empire" could be developed, the area had long remained a semidesert. By the 1930s, its population had fallen by 40 percent. Here and there ruins of farms with their rusty, wind-driven pumps and dry wells stood as witnesses of the futility of human endeavors. In 1929, the year of the stock market crash, the region was hit by a particularly severe drought. Mile-wide clouds of dust, swept up from the ground, ranged over the plateau, blocking the sun's rays.

Nevertheless, for a few years already, groups of citizens had been meeting to study means of irrigating the plateau. For example, the "Columbia Basin Irrigation League" advocated construction of a canal between Pend Oreille Lake in Idaho and the rainless plateau. Water would flow under gravity into the Spokane River and power turbines already in place. The powerful Washington Water Power of Spokane would strengthen its monopoly. A coalition of businessmen supported this project, which seemed to be gaining favor over the construction of a dam.

However, the cities on the west side of the Cascades often ran short of electricity in periods of drought, when the streams ran low. It also seemed to be a good idea to ensure a steady supply of power to public utilities.

The tug-of-war between supporters of the dam and those of the irrigation canal rapidly transcended the realm of civil engineering. Many small farmers and business people saw it as a struggle between public services and unbridled free enterprise; the people rose against the elite. Building a dam was seen as the state harnessing nature for the benefit of the common people. Modern technology held a double promise: shaping not only nature, but society as well. In the Soviet Union, Lenin had proclaimed: "Communism is the power of the Soviets plus electrification."

Irrigation would counterbalance the power of immense ranches and big enterprise by freeing agriculture. Small farmers would also be able to mechanize their operations and acquire electrical appliances, while small and medium businesses would set up shop in towns neighboring the hydroelectric power plants. Pollution and congestion in the big cities would be reduced, thanks to the relocation of those industries.

The vast and desolate Columbia plateau would become a Garden of Eden by building the Grand Coulee dam, strategically located in the heart of a region spreading over five states, with 3.5 million inhabitants; 3 percent of the U.S. population at the time.

A 100-mile-long (161 km) lake would be created upstream of the dam (Lake Roosevelt). Enough electricity would be generated to supply many large cities; the irrigated lands would grow cereals, vegetables, fruits and cattle to feed city dwellers.

Of course, some of these cities did not yet exist, which led some to think of the project as idiotic. In 1937, a newspaperman talked of the dam as "a gigantic white elephant, nearly as useful as the pyramids."

However, in 1934, president Roosevelt had asked the crowd during a visit to Grand Coulee:

What would you like? A great dam or a small dam?
"The great dam!" roared the crowd.
Then you'll have it…

Roosevelt was a man of his word and the great dam, seen here as it appears today, is a marvel of 20th century engineering. On the lower right a worker is seen carrying a 193-pound (87.5-kg) nut and bolt used to join a section of the generator shaft (photo taken in 1942).

Chapter 14

The Blessings of the Electricity Fairy

Work on the Grand Coulee began in 1937. The dam's museum overwhelms the visitor with statistics. Just a few will give an idea of the magnitude of this giant. At the bottom, it is 5,000 feet (1,500 m) thick, narrowing to 50 feet (15 m) at the top, the width of the road that runs across it. It stretches 5,223 feet (1,592 m) across the canyon. Seven thousand workers were hired; they struggled through the mud, the cold and the discomfort of their quarters. They fought thirst in the summer by swallowing salt pills and liters of beer. There were many injuries and 70 fatalities. Many were buried in the cement of the dam, victims sacrificed to the gods of hydro-electricity.

Sometimes, as a joke, the workers would slip sausages into the fingers of rubber gloves, making a hand that would seem to reach out of the fresh concrete. They would then call the foreman. "Everyone had a good laugh," said Ed.

There were few diversions. The workers would organize black-widow fights, or look for ways to keep rattlesnakes out. They also drank, fought and gambled away their pay. Soon the worksite attracted professional gamblers, petty crooks, drug peddlers, pimps and prostitutes. It also attracted poor families, ready to work hard to start on a new life, thanks to the dam.

Grand Coulee started producing electricity eight months before Pearl Harbor, in 1941. By 1943 the Columbia dams generated 96 percent of the electricity needed for the war effort. Aluminum smelters were among the heaviest users. Aluminum was a premium metal for the Boeing factories, which built 10,000 bombers during the war. The Portland and Vancouver, Washington shipyards were also major energy consumers. So was a mysterious establishment, Hanford, located in a deserted area near the Oregon border. Its function became better understood after the U.S. nuclear attack on Japan. The Hanford Nuclear Reservation, part of the Manhattan

Project, was home of the B-reactor, where the plutonium was produced to make the "Fat Man" atomic bomb used on Nagasaki, Japan on August 9, 1945.

The war effort had relegated the irrigation projects to the back of the stage, and it was soon time again to think about the more unfortunate citizens. The songs that Woody Guthrie had composed, on behalf of the Bonneville Administration to promote the construction of dams, became classics. For example, *Roll on, Columbia, Grand Coulee Dam* and *Way up in that Northwest*. In *Pastures of Plenty*, Guthrie sang:

> Green pastures of plenty from dry desert ground
> From the Grand Coulee dam where the water runs down
> Every state in this Union us migrants have been
> We work in your fight & we'll fight till we win.

Ten years later, would the Columbia River Project finally fulfill its promises? It seemed appropriate to launch the renewal of the Columbia Plateau development work with a public relations campaign worthy of its scale. Hu Blonk, a journalist, was called upon. As a publicity stunt, he decided to transform a desert lot into a working farm in a single day. The farm would be given to the country's most meritorious veteran. The Veterans Association, through its many chapters, contacted its members. Donald Dunn, age 31, was chosen from 20,000 possible candidates.

After serving as a tank driver, Dunn had returned to his Kansas farm after the war. Five years later, he lost everything through a major flood. Taken in by family members in Yakima, Washington, he had found employment as a farm machinery salesman. Six days before the birth of his third child, he witnessed the creation of a farm in 24 hours on an 86-acre (35-hectare) patch of sagebrush.

The project started at midnight, May 25, 1952 with fireworks. Eighteen power shovels and 50 tractors dug up the ground and spread the seeds. Five hundred workers erected the buildings. In this fairy tale scenario, the Dunn family was also given a cat and a dog as pets and spent their first night under their new butterfly-wing-shaped roof.

The next day, the whole country read about Farm-in-a-Day.

The Farm-in-a-Day house being constructed for Donald Dunn and his family. The photo was taken in May 1952 (U.S. Bureau of Reclamation).

Donald Dunn figured, shovel in hand, on page one of the newspapers.[1]

Dunn rented an additional 75 acres and seeded it with grass. He soon discovered that he was in no position to run a dairy farm, since he lacked the equipment required to treat and store milk. Transport costs also increased his costs. It cost him more to grow potatoes than he could buy them in town.

Small farms were doomed to failure—irrigation alone was not enough to ensure their success. Deep in debt, Dunn sold his farm. It was owned in turn by three others, all equally unsuccessful. It only became economical once it was incorporated into a much bigger

---

1. A description of the events and photos of the house and of Donald Dunn may be found on the Online Encyclopedia of Washington State website: < http://www.historylink.org/essays/output.cfm?file_id=8114>.

agro-enterprise operating over 450 acres. Dunn's farmhouse, with its futuristic roof, can still be seen nowadays near Moses Lake. A reporter who wrote about this event commented in glowing terms, "It was truly possible to create Utopia in a single day."

We were now overlooking the dark waters of the reservoir, apparently bottomless. I had the impression of looking down at a tomb. The surface of the reservoir looked like dull marble. I have visited Grand Coulee many times. Always, even in the summer under a bright sun, its deep waters retain their dark blue tone. Pleasure craft criss-crossed the surface.

At night, spectators can sit in the grass and watch a sound and light show. Laser projectors illustrate the history of Grand Coulee on the dam itself, used as a giant screen.

Nevertheless, whatever the season, I always experience an uneasy feeling in the vicinity of the hydroelectric marvel.

Ed's voice brought me back to reality:

Following the Columbia River, we relive the history of the world, as it took place in this part of the world. The Earth lives, throbs, breathes. It rises, spews fire, spreads ice. Then it calms down. Erosion slowly smooths it out. Hills and mountains gradually wear down; leisurely, rivers develop new courses. Then comes another era; the Earth heaves in a new wave of unrest and pushes the Columbia off course. But take note: to this day, the Columbia is still there. The changes are part of its existence. Each upset forces it to alter its course, open a new way and dig its bed anew. That is how the river persists. And if you have looked carefully at the fluctuations it experienced in the past, you will readily imagine what awaits it in the future.

However, when you stare at the Grand Coulee Dam, you are forced to admit that there is one factor that can compete with geology—the work of man. For better or for worse.

I had already spent a long time looking at books, magazines and films to become better acquainted with North America, but now I began to appreciate the prehistory of that corner of the world. I attempted to grasp the factors that influence the scenery—long term causes (volcanoes, glaciers, earthquakes, erosion) as well as accidental events (forest fires, landslides), not to forget the work of man. For better and often for worse, no doubt.

Chapter 15

The Electric Garden

We were now back on the road, beyond the mesas surrounding the Columbia, and approaching the rolling plateau. The road stretched straight as a ribbon over hill and dell. Alfalfa fields reached to the horizon. At 60 miles (97 km) an hour, the wheels and the transmission shook alarmingly. Our car gained little advantage from speeding down the slopes; little of its momentum was left when reaching the top of the hills.

Tall, metallic silos rose against the horizon: the village of Wilbur. Ed parked in front of a store where he stopped to buy milk, taking his good time. We entered a café. Ed read the newspaper. Outside, it was getting colder and colder. On the way out, Ed looked at the sky and said, "We should turn back."

I asked to drive. Ed agreed. It started snowing at Grand Coulee—a light dusting that lasted until Nespalem. As we approached the forest and the curves of the Disautel Pass, the snow became much heavier. I no longer dared shifting into third gear; we slowed down to 25 mph (40 km/h). It was getting dark and the visibility dropped to 70 feet (20 m). Sometimes we slipped sideways, crab-like.

"Ed, don't you want the wheel back?"
"Gee Whiz! We can't stop! We'd never be able to start again."

With great precaution, I drove in the middle of the road because I couldn't tell where the pavement ended and where the ditch began. Suddenly we met a car. I steered to the right, nervously. We skidded and then quickly recovered.

"The important thing is to stay on the road," said Ed.

We met only two other cars. All I could see through the windshield was a white screen of heavy snow. On both sides, I could make out the black mass of fir trees. I feared that some animal—a deer, a bear or a coyote—might cross the road. A sudden move or a collision with an animal would surely put us into the ditch. And what if a large and mysterious creature suddenly appeared in the headlights? The idea of an encounter with a sasquatch briefly crossed my mind. I quickly dismissed it and concentrated on my driving. I had the spooky feeling that the forest was trying to suck us in. I kept to the middle of the road, as far from the trees as possible, to escape the lure of the dark forest.

Ed must have been reading my mind. Perhaps it was enough to see my hands clenched on the steering wheel. "Don't let Mother Nature lead you astray," he said.

Nearing the Okanogan River, the road descends steeply into the valley. The snow cleared up. The streets of Omak, deserted on a Sunday night, were covered by a thin layer of fresh, powdery snow. A few minutes later we reached Ed's home. I sighed. Ed gave me an amused look. He busied himself with the stove. I poured some bourbon, sat on an old couch and wrote in my journal.

Sunday, February 20, 1994

Between opposite shores of the dazzling Columbia, seen from the road above, there is a sharp contrast; on one side, light-colored granite, on the other black basalt. Black lava dikes run through the granite, filling gaps left by volcanic fractures. Sometimes the dikes contain cavities lined with crystals, sometimes precious gems. The Earth is active with volcanic activity. Fire gives birth to precious stones.

I have the feeling that here the Earth speaks and is heard by the people who walk it. To draw an analogy to a tape recorder, these people are the reading heads of the Earth.

Our plates on our knees, we ate while watching television. My journal records the evening's shows. First, a documentary discussed the plight of American soldiers suffering from the sequels of the "Desert Storm" campaign. During the war with Iraq, they had manipulated spent-uranium, armor-piercing shells without having been warned about their nature. This was followed by a program about rape, which ended with the question: should the guilty be castrated? Finally, for dessert, we were treated to an application of behaviorist science known as "deportment psychology." In a jail, pictures of children and adolescents were projected to an audience of sex maniacs while these inmates' degree of sexual excitation was being measured. It was hoped that by a suitable treatment or conditioning, they could be cured, at least temporarily, for a permanent cure would require repeated reinforcements.

Ed responded to each show with appropriate expletives. As I watched the various shows, I experienced a mixture of embarrassment, anger and disgust. I was about to retire for the night when Ed said he had something to show me. Curiosity took over and I followed him up the steps to the second floor.

He opened the door onto a wall of greenery: 80 tomato plants climbing up six feet (1.8 m) on supports. Industrial size bulbs, as big as children's balloons, lit the room, their light reflected by four-foot-wide (1.2 m) aluminum shades.

In a corner a tap fed a water barrel. A floater valve controlled the level and a pipe ran from half-way up the barrel to irrigate the tomato plants. Ed explained:

> This way, I harvest tomatoes all year round. They are now rather small because my plants are too close together. The room is lit permanently. A timing mechanism turns the lights on automatically from midnight to 6:00 a.m. If I were not honest, I could easily engage in some illegal gardening. So, what do you think of this "good old American know-how?"

Ed loved that expression and used it frequently. There is no denying that he took childish pride in a creative and inventive America. I often had the opportunity to admire the results of my friend's fertile imagination. Some rather debatable projects never came to fruition, probably for the best. I was often the beneficiary of this famous inventiveness: there seemed to be no tool that Ed couldn't take advantage of, no technical difficulty would deter him, no obstacle ever set him back. He continued:

> Four times a second…The computers at the Grand Coulee Dam compare the rate of electrical generation with the needs of the consumers. When I turn my lights on, the computers command a slight opening of the sluice-gates. We

have as much electricity as we wish and the Pacific Northwest enjoys the lowest rates in the whole country.

As I dozed off in bed, I thought of the "electrical garden," of all the benefits brought by the dams to the small rural communities, about the growth of orchards and vineyards, and of the herds of cattle. The dams had brought the desert back to its earlier status as a Garden of Eden, in the days before the rise of the Cascades; a wonderful instance of the reconquest of nature, a utopia of regeneration.

But I was also thinking about Robert Pyle's warning, deeply etched in my memory:

> "most of all, Bigfoot shows what could have been and what still could be, if only we treated the land as if it were really there. For the very wildness from which the Bigfoot myth emanates is disappearing fast...when the topography is finally tamed outright, no one will anymore imagine that giants are abroad in the land."[1]

---

1. Robert Pyle, *Where Bigfoot Walks*, p. 17.

Chapter 16

Good-bye For Now

I had just walked up to a second-floor room with Ed. He had lifted a sliding window to fill one of the bird feeders hanging from branches attached to the wall. At the sound of the sliding window, a flock of grosbeaks rushed to the feast of sunflower seeds. After this morning routine, Ed went on with his day.

On that cold Monday morning, I decided to bask in the warmth of the wood stove and to indulge in a second cup of weak coffee. Leafing through a collection of old clippings from local newspapers, I came upon a bold title: *Through the Okanogan County, Grandeur and Beauty of that Region Described,* subtitled: *Rich in Precious Minerals, Fertile Lands and Wide Pastures, It is Being Developed Faster Than Any Other Portion of the United States.*

According to that article, written in 1892, Okanogan County had remained until recently "terra incognita." It was over 120 miles (200 km) long and of similar width, bordered in the east by Stevens County, in the west by the Cascade Mountains, in the north by the province of British Columbia (Canada) and in the south by the Columbia River.

It was said to contain the whole gamut of natural resources found in the various parts of the United States. No other region could match its perfect climate. "It is by itself a whole empire," claimed the article, which continued by mentioning that the eastern part of the county belonged to the Colville Indians, but that it was expected that the president of the United States would soon open up the northern part of the reservation to miners and settlers. Thousands of homesteads would be established and the exploitation of the mineral resources would bring millions of dollars to the nation's coffers. Because of the absence of rail connection, the county had long remained isolated, but soon one or more roads would allow the rest of the world to trade with this "New Eldorado."

Actually, in the northeast part of Okanogan County, in the Lake Chelan area, one finds gold, silver, copper and lead deposits. The Methow River valley contains good arable lands as well as high-quality coal deposits. Further north, in the villages of Ruby City and

Conconully, one reaches the silver belt. Still further north, closer to the British Columbia border, in Loomis and Oroville, one is in the gold belt. In 1892, the Indian reservation was described as "The Land of Canaan, rich in mineral resources, where run the milk and honey."[1]

By settling in Riverside, Ed Fuchs had made a wise choice for the location of his prospecting supplies store. He was 15 minutes away from Conconully and one hour from Oroville by road. However, in the winter, the clients were few and far between; because of the harshness of the weather there was hardly any prospecting. Ed took advantage of the season to make repairs preparatory to the grand opening of the store in the spring.

Traditionally built schools in Great Britain as well as in the United States include a large hall, sufficient to assemble the whole student body. That room includes a stage and a podium from which the principal speaks to the pupils. It hosts feasts, dances and graduation ceremonies. It is surrounded by a gallery, held by pillars, which can accommodate an overflow of spectators. In Ed's school, the assembly hall had also been used as a basketball court. It was a very large room, as high as a two-story building; the roof was (and is) also the ceiling of the assembly hall.

Ed had transformed that room into a store. He built under the gallery a series of stalls, about 10 by 16 feet (3 x 5 m) in area, separated from each other by a wall of wooden boards. Half a dozen of these line up each side of the hall. One was to be devoted to literature: books on geology, prospecting, analysis of minerals; magazines, maps, monographs. Another included metal detectors. A third, diving gear—including wet suits and diving helmets—for prospectors who might have to explore streams or lakes.

In the middle of the hall, lay a collection of bulkier machines: floating dredges, pumps, conveyor belts, and aspirators. All the treasures in this cave of Ali Baba were exposed to drips from the

---

1. Hilderbrand, 1991. p. 438

leaky roof. The cracks in the roof had to be filled, from the inside as well as the outside. Ed had hired for the purpose two young workers, Russ and Don. One of them was busy on the roof while the other, inside the store, was standing on a tall ladder, nailing to the ceiling the boards and joists that Ed was cutting to length.

As usual, Ed worked without interruption from 10:00 a.m. to 7:00 p.m. He nevertheless allowed his employees to break for lunch and for a smoke now and then. That night Russ, the carpenter, was about to leave. I invited him to join me at the Riverside Saloon, the only liquor outlet in the village. There were a few clients—the 21st of February is a holiday, President's Day, in honor of George Washington.

Russ was a man of few words. He was of course tired from a day's work, but that explained only in part his stressed look. He was about 25, tall and dark, trim and robust. He came from Northern California. His father had left home as Russ approached 14. At 17, without any degree or diploma, he set out for the big city, which he left after a few years. By the tone of his voice, I seemed to understand that he had fled some great danger or some dangerous habit—alcohol, drugs?—and had wandered northwards. In Okanogan County the low cost of living compensated for the salaries, which were also much lower than in the city. Life was simpler, far from the temptations and the chaos of the city. Russ had settled in.

He became animated when telling me about an incident from his lonely childhood. One day he had found a wild cat whose mother had died from a hunting wound. The animal was his faithful companion for years, obeying only him.

Russ now lived a few miles from Riverside, on the slopes of Mount Tunk—altitude 10,000 feet (3,500 m). He asked me why I was visiting Washington State. He listened carefully and quietly, watching me. Finally he made up his mind to talk:

> It was in January 1993. The frozen pond was thickly covered with snow, about three feet (1 m) thick. I live in a tepee, which I fashioned from sheets of mylar, a thick and sturdy plastic. To reach it, one must park the car and continue on foot up a trail for a few hundred yards. On the right, there is a thicket, then a fallen tree, then the pond. The trail bends towards a viewpoint, atop a steep cliff; that's where I

set up my tepee. The view is superb. On the left, there is another thicket.

One night, walking home with my flashlight in hand, I noticed some large footprints in the snow, rounding the pond and going between the two thickets. The tracks disappeared into the second thicket. Nearing the tepee, I saw a tear in the wall, about chest height. Worried, I hurried inside, expecting a major plunder. There was an empty can of icing sugar on the ground. Its top had been shorn off as neatly as with a razor blade. Puzzled, I walked around the tepee. Near the tear there were tooth marks, similar to a denture imprint. But what a denture! Much larger than that of a horse or a cow. I am very familiar with these as I often take care of a neighboring farmer's animals. I put some shoe shine on the edges of the tooth marks and pressed them onto cardboard. Here they are.

A glance at the shape and size of the tooth marks showed that these were certainly not human teeth! Russ continued:

I walked back out towards the footprints. Unthinkingly, I stepped into them. Each footstep was more than five feet long, and that in thick freshly fallen snow. Naturally, I altered the original prints so that it was no longer worth trying to preserve them, but it seems to me that they were one and a half times as long as my boots.

What else can I say? I usually avoid talking about this strange incident. But one day I met a professor, a doctor in zoology, who lives nearby. She took me seriously. You could phone her.

It was now time for Russ to leave and to climb up "his" mountain. He didn't go out much, he said. In the summer, after a hike in the steep hills, he liked to sit at the viewpoint, his feet dangling over the edge, gazing at the stars twinkling in the waters of the Okanogan River.

The following morning, at nine o'clock, I boarded the bus in front of the old Caribou Hotel in Okanogan. As far as Wenatchee, there were only two passengers: myself and a young woman, about

35 years of age. Her name was Becky and she prospected for precious stones, mainly garnets found in granite and gneiss around Coeur d'Alene, Idaho. She would load her truck with rocks and take them to a small craft shop where they extracted the garnets. She opened her heavy bag and showed me samples of her rocks—rather ordinary stones, from which gems would emerge after the right treatment. Becky led a tough life, poor in comforts, but she had chosen privacy and freedom. She told me she saw a cougar recently—"a rare and marvelous encounter," she said.

Just before Stevens Pass, the driver asked us to help him put chains on the rear tires. The snowbanks were five feet (1.5 m) high. Over here, a car was overturned; over there, there was one in the ditch. The bus reached Seattle an hour late, at 5:00 p.m.

I had now taken leave of the Columbia River, of the Okanogan Plateau, and of my friend Ed. I was going to spend a few weeks in Seattle, reading in the University of Washington library.

But I had still not completely turned the page of the Great Book. My memories of those four days in February remained vivid. I wrote in my journal:

Thursday, March 10, 1994

It is night. I am thinking of that trip as far as Wilbur. Suddenly, I perceive a different significance: it was a voyage to the depths of the great dam, dark refuge, and reservoir of disturbing mysteries.

Before Grand Coulee, one sees towering rock faces along the road, shrouded in mystery, announcing the proximity of the great dark mass.

We left the desert to reach the forest, followed by prairie, with Indians and their horses.

Reaching Wilbur, we had to turn back to loop the loop as quickly as possible. The warning was given by the lake; the heavy snow created the aura of mystery, the invisible world inhabited by deer, foxes and the giants of the forest.

It was a good thing there were two of us. Two imperfect friends, the one anxious, the other not—or at least, less so—unsure as to their goal, but driven to return home. That night, my eyes were sore. What was I seeing, really?

Chapter 17

The Dalles

It is easy to lose track of time in the quietude of a library, especially in the Special Collections hall. One's bag and coat are left at the door. A librarian hands you a number to place prominently on your desk. After a few minutes, someone brings you the books that you requested. You have pen and paper in hand for note taking. If the state of the document allows it, you may obtain photocopies within 24hours.

Ground-level windows light up the semi-subterranean room. When you lift your head, you notice that it is raining or that wet snow has begun to fall. It's winter in Seattle. The heating functions well, the lighting is perfect, the desk is wide enough to spread maps and atlases.

Your legs are going to sleep for not having left your seat in hours. You get up for a walk through the tree-lined campus, the home of gray squirrels. You might perhaps walk as far as the Burke Museum, where the main displays are devoted to American Indian cultures. There, you should keep in mind that the history of the Indians was not written in books, but carved in totems, masks and talking sticks; it has been embroidered in their blankets; it has been transmitted in this manner through symbols of honor and authority.

That day, I went to the basement of the Burke. I wanted to see the rhinoceros, a member of a now extinct species which inhabited the Pacific Northwest during the Miocene. The cement statue of the animal, accurately detailed, is part of a scene that recreates a drama of 15 million years ago. The image of the lava flows of the Columbia plateau, more massive than those that gave rise to the Hawaiian Islands, returns to my mind. For hours, the rhinoceros must have run away from the hot lava advancing over an endless front. Finally, the exhausted animal had been overtaken and buried into the lava. In contact with the animal's body, the lava had cooled, leaving a perfect mold. In a wheat field, geologists had discovered a large mysterious hole—the cavity left by the body of the rhino. They filled it with plaster and made a mold which was then used to cast the Burke Museum's cement statue.

Even in Seattle, the capital of aviation and information technology where tomorrow's world is being created, the wave of volcanic magma was catching up with me, as it had with the rhinoceros. Paradoxically, I was threatened with paralysis, hypnotized by the series of cataclysms that had deeply modified the environment. I was imagining those periods when life was returning to the disturbed lands, while the Columbia was patiently and continuously carving its new bed. These spectacular cycles, taking place on geological time scales, were becoming an annoying fixation. On the one hand, they familiarized me with the genesis of the area, but on the other, they were becoming an obsession, hiding from me phenomena occurring on a more human scale.

I was wrong to be so concerned. The variety of sources that I was discovering, the thoughtful comments from my friends, the welcoming attitude of all the investigators I met, as well as the generosity with which they shared their knowledge, were soon to dispel my worries.

Strange clues led the way towards a more human plane. Rather disquieting clues, such as a feat, found in a geology textbook, which I immediately took note of. In 1935, in the Hawaiian Islands, an eruption of the Mauna Loa volcano was threatening the town of Hilo. The American government sent out the 23d Bomber Squadron which dropped 600-pound (270-kg) bombs on the lava flow and managed to deflect it.

I had not yet fully understood how people succeed in diverting the course of nature. I was to become gradually more aware of it by coming closer, once more, to the *Nch'i-Wana,* the Great River, as it is called by the Sahaptin Indians who live on the middle Columbia.

Peter Byrne, director of the Bigfoot Research Project, was welcoming me to his headquarters, near the town of The Dalles. I opened up his book *The Search for Bigfoot* to find out more.

The Dalles is a town of some ten thousand people lying on a bend of the Columbia River, in northern Oregon, about eighty miles (129 km) east of the city of Portland. It was originally called Les Dalles. The word "dalles" is French and means stepping stones, or flagstones, and is presumably referred to the flag-like stones that lay across the bed of the Columbia River at this place and which can still be seen just below The Dalles dam today. The name was given to the area by early French settlers and traders working the Columbia River. History records that the first settler in The Dalles was a Frenchman named Joseph Lavendure. He arrived in 1847. He left again in 1848 and settled in California.

Byrne then goes on to state:

> Today the city is pleasant and quiet. The town has many old wooden houses that lend it a certain charm and dignity and the mighty Columbia, flowing quietly through the bend on which the town is situated, reminds one of the colorful past that is the history of northern Oregon. There are a few Indians in The Dalles now, but the tribe—the Celilo—that once lived on the bend of the river and fished its waters for salmon and sturgeon are almost all gone. They called the Columbia the Wauna, a beautiful word that is somehow more fitting than its present name. In years gone by the Celilo fished the Wauna with hand nets and salmon was a major part of their diet. The building of the dams at Bon-

Peter Byrne's Bigfoot Information Center in The Dalles, 1970s.

neville, Cascade Locks and The Dalles put an end to their livelihood on the river.[1]

The Dalles area is equally distant from British Columbia in the north and California in the south. Peter Byrne had chosen it for its strategic location. A few years ago, he had set up his Bigfoot Research Project in a nearby village, Clarksdale, at the foot of the fascinating snowcapped Mt Hood. Even during summer, there is snow on the mountain, although the temperature in the valleys can reach up to 100°F (38°C). The Cascade Mountains are nearby. Northwest of The Dalles, there are forests, fields, and verdant hillsides. The shores of the Columbia, however, belong to another world—arid and fringed with sharp rocky outcrops. Small bushes grow here and there as well as scattered oaks and resinous trees. In the summer, the yellowed grass waves in the wind, a wind which blows nearly permanently and attracts windsurfers. Experts from around the world flock here to practice their sport on the windy Columbia. In July and August, the nearby town of Hood City becomes a windsurfing mecca.

As one descends the river, its shores broaden. In 1805, explorer William Clarke wrote in his journal, *Ocean in view! O! The joy!* But alas, he was only at Pillar Rock, 16 miles (25 km) from the mouth of the river. William Dietrich described their predicament as follows:

> The white-capped waters of the estuary and the long carry of the sound of booming surf fooled them into thinking they had reached the Pacific. Heavy storms, incessant rain, and seemingly impenetrable dense forest kept them pinned to the riverbank, soaked and miserable, for ten days before Clark finally saw the true Pacific.[2]

The sheer magnitude of the entrance to the Columbia, and the sight of the dangerous bar that ships have to navigate to enter the river, are breathtaking. But overall the river now appears quiet, even

---

1. Peter Byrne, *The Search for Big Foot* (1975), pp. 66–67
2. William Dietrich, *Northwest Passage,* p. 100

lazy. It is only by going to museums, examining old prints and photographs and reading the accounts of travelers, that I got to know the river as it was before being domesticated. There are nowadays no less than 14 dams—and accompanying reservoirs—between the ocean and the river's source. The Columbia today is as smooth as an English butler.

It is probably easier to imagine the fury of yesterday's river when looking at The Dalles—or what little is left of that giant staircase—where clouds of foam spray the rocks which the Wishram Indians called The Bridge of the Gods:

The Mountain Spirits often quarreled with the Spirit of the River. During one of their confrontations, the Mountain Spirits erected a great stone barrier across the river and blocked its course. But the Spirit of the River, in its mighty strength attacked the barrier and dug a subterranean passageway. The river rushed through the tunnel, but the roof stood fast, becoming a bridge that allowed people to cross the river. It was called *Tamanawas,* the Bridge of the Gods.

U.S. Postage stamp commemorating the Lewis and Clark expedition. (Issued in 1954.)

Chapter 18

The Salmon — A Priceless Gift

Celilo Falls, at The Dalles, displayed the might of the untamed river, a 12-mile-long (19 km) series of basalt islets, rocks, rapids and narrows. Not as high as Niagara, it nevertheless was one of the largest waterfalls in North America. The Natives, perched on high wooden platforms, fished with woven cedar-bark dipnets. Others speared the fish from rocks overhanging the river. Many fishermen fell from the scaffolds, dragged down by the weight of the salmon in the nets, and drowned in the turbulent waters.

Salmon made up to 40 percent of the diet of the tribes living along the river. It was the Ancestral Creator's most precious gift to the Indians, conveyed by his emissary Coyote, the trickster. A prophetic Wishram myth tells the story of two sisters who had imprisoned the salmon in a lake made by diverting the river. Hearing this, Coyote transformed himself into a newborn baby. He set himself in a basket adrift on the water and started crying. The sisters were gathering driftwood; they heard the cries of the baby, rescued him and took him home. Later in the day, they left home to gather roots with their digging sticks.

Left alone, Coyote recovered his original form and worked at destroying the dam separating the lake from the river. On the fifth day, one of the sisters broke her digging stick—a bad omen. Anxious, the two women hastened home and discovered Coyote

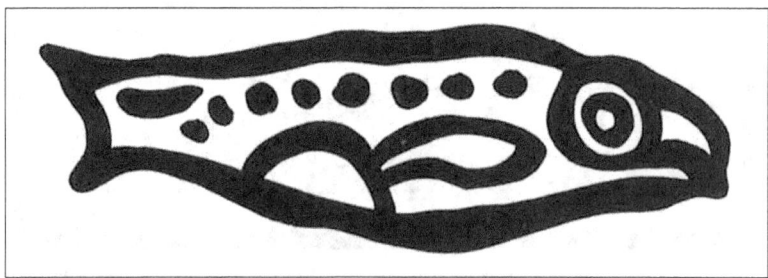

Stylized salmon, after the North American Indian tradition.

destroying the dam. Too late, they rushed to stop him: the earth dam collapsed, the lake waters poured into the river and the salmon escaped.

Coyote lectured the sisters: "By what right are you keeping the salmon for yourselves? Soon, people will fill the world and salmon will be their food." Coyote changed the two women into sparrows, which gather every year on the shores during the fishing season.

The degree of refinement of geographical nomenclature in Indian languages reflects the cultural and economic importance of a region. At Celilo Falls, each rock or islet had its own name. For example, *Swaycas,* meaning "long pole," was the rock were a 20-foot-long (6 m) pole ending in a dipnet was needed to catch salmon. A rock was named *atiim* (the roar of the falls) because it was immediately next to them. Flat rocks over a clear water area were called *tayxaytpama* (good for harpooning)—a spot suitable for that type of fishing.

Indian fisherman at Celilo Falls.

Each location had its own importance, depending on the seasonal water level. Each family would erect its own fishing scaffolds, which could be used by others by permission of the owners, who felt they had to share their wealth with relatives and strangers. Fish were given to elders who came to watch; a foreigner would be invited to catch a fish.

Indian place names yield valuable information on traditional ecological knowledge; so does the vocabulary associated with such a basic activity as fishing. They had at their disposal a linguistic tool, refined over centuries, co-evolved with improvements of their physical tools and fishing techniques.

The Mid-Columbia tribes spent their time gathering, hunting and fishing, following the rhythm of the salmon migrations. Other species were harvested using selective fishing or trapping tools, using for example nets of the right mesh size, or plants best adapted to the purpose. The Indians were well versed in the properties and use of the natural materials at their disposal. The Columbia and

Yakima Indians were well aware, for example, of the bactericide properties of caluks, a kind of wild carrot *(Lomatium dissectum)*.[1] They crushed its roots into a pulp for external use. They also used it in compresses to cure saddle sores on horses. At the first sign of a cold, one drank an infusion of the plant. It could also be used as a concentrate in a hair rinse to get rid of dandruff. Lomatium also acted as a poison for some fish which it momentarily paralyzed. When poured into a back eddy or in a quiet reach of a stream, the plant had time to achieve its effect without permanently endangering the fish. Only what was needed was harvested.

However, the salmon, inexhaustible when the Indians fished it, became endangered when the European colonists began to exploit it. From that time it faced, as did the prairie buffalo, the risk of extinction. William Dietrich writes that as early as 1807 the future of the salmon was in danger. At that time, Napoleon Bonaparte was looking for a new and safer way of provisioning his armies in the field. Attracted by a 12,000-franc reward offered by the emperor, Nicolas Appert, a Parisian confectioner, initiated research that led to the development of canned foods.

As early as 1829 there was an attempt to start a cannery on the Columbia. It was not until 1866, however, that canning really took off. That year, the first cannery put into tins 144 tons of King Salmon (Chinook)—the great rush had begun.

The Impossible Return

The salmon's life cycle is a subject of universal fascination to all, whether they fish or not. The image of the salmon is etched in our memory; its name evokes a large silvery fish leaping a rapid or a waterfall.

A few years ago, I had the opportunity of coming close to this fish on Kodiak Island, in Alaska. It was August; the sun was barely piercing the clouds and the diffused light resembled that of a March day in our latitudes.[2] I was standing about 500 yards (460 m) from

---

1. *Lomatium dissectum* is sometimes called biscuit root, or Indian carrot. It was used primarily for upper respiratory infections. Today, naturopathic physicians use it in the treatment of viral diseases.

the mouth of a small turbulent river. Off to my right a sign proclaimed: "No fishing beyond this point."

At that time of year, the salmon upstream migration was in full swing, and the spawning grounds were protected by the Alaska Department of Fish and Wildlife. Across the bay I could see the dark shape of the mountains against the gray sky. Small waves broke against the pebbles of the shore. Although I was some distance from the mouth of the river, I did not have to wait long. Soon, my line was pulled taut; I let my reel unwind. Gradually, I slowed down the unwinding, and then stopped it altogether. I then embarked on the long process of tiring my catch and of bringing it, step by step, to the shore.

Finally, it was no longer struggling; seeing its belly tilted at 45 degrees towards the surface I feared that it had died of exhaustion. I pulled the fish onto the pebbles and took out the hook. I then placed it in the water, head pointing away from shore. I massaged it to rekindle any strength left in its powerful streamlined body. It took a while, but its breathing became more regular, and the body twisted back to its normal attitude, belly down. The air temperature was no more than 51°F (12°C) and the icy water was making my fingers numb. I could hardly feel them, stiffly cupped, sliding back and forth down the body from gills to tail.

Suddenly, I was holding an empty form: the salmon had sprung forward, like an arrow. Today I can still feel that my hands have preserved the memory of that form.

I continued fishing for some time, easily bringing in smaller catches. One of them had on its side a gaping wound, probably caused by the bite of a seal. Salmon are very hardy; they withstand minor injuries and the bite of fishhooks.

Between June and December, depending on the region, from Oregon to Alaska, salmon leave the ocean to swim up rivers and streams. They travel in schools, making the fortune of lucky or clever fishermen, working nowadays from powerful seiners pulling nets as long as 1,000 feet (300 m). When they see the fish leaping out of the water, the lookout announces their presence by shouting: "Jumpers!!!" That's what Caesar's soldiers called them when they

---

2. The author lives near Nantes, France (latitude 47°N); Kodiak's latitude is 57°N.

first saw them on the Atlantic shores of Gaul: *Salmo,* from the latin verb *salio,* to jump. But there is no real need to climb up the mast to detect them—modern means like sonar and sea planes can do it too.

Salmon delay their entrance into fresh water, sometimes staying up to a month in the estuary. They have to adapt to the passage from salt to fresh water in a process that reverses what they had to do when, as juveniles, they entered the ocean—modifying gills and kidneys to adapt to a more saline medium. They now have to readapt to swim upstream, sometimes for hundred of miles, to the gravel beds where they hatched. Thanks to their sense of smell, salmon can home-in on their natal stream. Sometimes they wander off into a sidestream, turn back, and continue their search. To quote biologist Alfred L. Meister:

> They have only one function, for which they are biologically programmed and that is reproduction, the survival of the species. They have just enough fat stored in their system from their years at sea to sustain them, provided their trip is not interrupted. And there is little else on their minds except to get where they know they must go.[3]

As soon as she arrives on the spawning grounds, the female digs a nest—a redd—with her tail. She deposits her eggs at the bottom and the male then fertilizes them with his sperm.

For all Pacific salmon, as well as for most Atlantics, there is only one spawning. After covering her redd with gravel, the female and her mate, exhausted, drift downstream. When they entered fresh water, their color changed, shifting from the silver to spreading patches of red and green. Some spent spawners in their scarlet livery try once again to buck the stream; their strength is quickly gone. An emotional observer, such as I was, might in a ridiculous gesture even attempt to assist a salmon upstream. All in vain, of course; like the others of his cohort it will soon lie dead on the shore. Bears, wolves, coyotes, cougars, raccoons, ravens and sea birds will feed on its flesh; the remains will enrich the waters of the stream.

---

3. Alfred L. Meister, as quoted in *Salmon* by Atsushi Sakurai, p. 5.

Strength and beauty of the salmon.

"Nowhere else in nature is there such a massive and simultaneous expiration," wrote John N. Cole.[4]

Over the winter, the eggs gradually mature, as long as the redd is not disturbed by a sudden flood provoked by the opening of the gates of a dam. However, other perils face the some 3,000 eggs lying in the redd—chemical effluents from factories, agricultural fertilizers, and smothering by the sediments eroding from the riverbanks (often as a result of logging).

In the spring, the eggs hatch and the alevins leave the gravel. A few months later, perhaps a hundred fry which escaped predators start on their journey to the ocean. Full of life, only five inches long, these juveniles are already leaping—miniature versions of the adults they will become if the "Salmon Spirit" grants them life.

Our young salmon still has a long voyage before reaching the ocean; it must also go through a major physiological transformation before entering salt water. Human obstacles add difficulty to the migration.

Irrigation channels lower the level and slow down the flow of the river; young salmon may get trapped in those side channels. Most manage to find their way downstream. Approaching the first reservoir, the river slows down, muddy bottoms replace rocky ones,

---

4. John N. Cole, as quoted in *Salmon*, by Atsushi Sakurai, p. 7.

and algae hide predators. Soon the flow accelerates near the dam. The young salmon is sucked in, falls 50–75 feet (15–23 m) through a dark turbulent pipe, and is finally spun through the turbines to emerge disoriented in a whirlpool of bubbles. It has been estimated that passing through eight dams leads to a 97 percent loss of young salmon.

With any luck, young Salmo reaches the lower Columbia, below the Bonneville dam. By then it is weak; food is scarce because the river bed has been sterilized by the dams and the dikes that border it. Here and there, as at The Dalles, rocks and islands have been dynamited away.

The wild pre-electrification Columbia did not treat the salmon so harshly. The tamed Columbia, placid and calm, without its falls and rapids has multiplied the perils of migration. When, after three years at sea, the adult salmon returns to reproduce, it must escalate fish ladders to bypass the dams and return to its original spawning grounds.

We must admire and respect those fish that succeed to reach their goal. To quote John N.Cole again: "Contact with one of these is evidence of immortality."

The Ichtyoid: A Fantasy—Perhaps?

A few months had passed since my visit to Alaska and my encounter with Salmo. For some time now I had been renting a lock-keeper's house, thick-walled and ancient—built in 1807—set in the midst of a network of canals in Saint Etienne de Montluc, near Nantes, in Brittany. In the early morning I often saw a heron fishing near my door. In the evening, I baited bundles of vine cuttings and lowered them with a string into the canal. Eels came and nestled into them; a dinner delicacy. At night, I lit old-fashioned oil lamps.

It was Easter. I spent three days reminiscing about the great fish I met in Alaska, an unforgettable encounter. I hesitated for a long time before including here a science fiction story that sprung to my imagination. I finally decided that it would be useful, and would help to better understand the fate of Salmo.

The Ichtyoid

First, the man had dived—into the sea, into a river? He wasn't sure—his head enclosed in a transparent helmet. At the shallowest level, he ran into some kind of algae that stuck to his helmet, which reddened at their contact. The algae slipped off.

Head first, falling like an anchor, he continued downwards only to be stopped by blue algae. This time the helmet shifted from transparent to milky before the vegetable straps released their hold. He came back up slowly, like an aquatic yo-yo.

He was puzzled. Should he think of the algae's slow envelopment as an aggressive move? They rolled up gracefully around his helmet, as so many tongues caressing his skin. Had the catalytic luminous helmet not functioned, he might perhaps have fallen victim to the insidious torpor: another diver caught in the net of predatory algae. He imagined himself overcome by a final and lethal sleep, his intact body gently bobbing in the deep currents. It was hard to believe that death could creep up in such a quiet and comfortable way. The absence of any blow made the danger unreal.

The diving suit had been fashioned to protect him against known dangers, like sharks, manta rays or moray eels. But no one had thought of algae as dangerous. His recent experience has shown otherwise, yet he could hardly believe it. He wondered if anyone would believe him. He could already hear the comments: "Rapture of the deep! Seeing the algae beginning to dance is a clear sign of nitrogen narcosis."

He swam back to shore, drew himself up and stepped on the mossy floor. The broken twigs underfoot were too rotten to creak. Silent as a shadow, he walked slowly away from the dark waters. Soon he was under the canopy of the trees; a maze of spider webs stretched between the branches—a thick network of blue-gray threads which damped drafts and muffled sounds.

He stopped, his helmet under his arm, a dark silhouette

on a gray background. A powerful beam of light shined through the dark and somber sky. The tree trunks glimmered. The forest sucked in his thoughts, entangling them in the network of branches and spider webs; neither disappearing nor dissolving, they thickened the skein of the forest.

He compared himself to a skyscraper in the pea soup of a fall morning, upper floors in the light, lower levels bathing in the fog. At that moment he became a floating head, its body hidden in the fog like the foot of the skyscraper. His brain did not throw light on the situation but continued to simmer with vague mental activity. Smiling, he thought, "I'm just like a hurricane lamp."

Snapping his finger, he tapped the surface of the helmet with his nail. The sharp click hung in the air like the glow of the skyscraper of his imagination. He needed that sharp noise to tear himself away from his reverie, to avoid spilling the puny contents of his small brain into the unfathomable bowl of the forest—on the brink of losing himself entirely.

A word came to his lips: *psychopump,* meaning, "The forest takes and keeps; it never gives."

The unchanging color, everywhere the same in spite of small variations, obscuring shapes and details. No noises, no reference points. A formless, soundless, disorienting world: a vertigo which readily turns into torpor. Danger!

Motionless, detached, paralyzed, as in last night's dream. But was it a dream or a memory? The harbinger of future events or the reflection of a parallel life? Standing at the edge of the forest, he had just relived his dream. He drummed his fingers on the helmet. Time to pull out of this dream!

When he pressed the door-opening button he remembered why he was there. He and a few others were responsible for the success of a lengthy and expensive program.

As he opened the door, he heard the words of a song: *Then she clearly understood if he was Fire, Oh, then she must be Wood.*

Sitting in a corner of the common room, Jim reached

out and turned the music off, adding, *Then the flames enveloped Joan of Arc*. He dog-eared the page of the book he was reading and closed it.

The diver in his dark suit looked at the title: *Magic: the Cryptic Art of Life and Death*.

Jim was the seaplane pilot. One could readily recognize him from afar by the bright red felt hat he always wore—some kind of humorous trademark. The book was something else again.

Jim said, "I flew over Eagle Pass. I saw three or four leaping but they are still some way off. There are small schools approaching the coast, a good sign that they'll be here soon. As usual, the others are in the lab. After the beatniks, the computerniks."

The whole team was now gathered around the big table in the kitchen-dining room. In a corner a small light showed that the microwave was powered. Without any good reason, he always sat as far away from it as possible. They were listening to old-time tunes—stories of hobos, loggers, scorned lovers, and women in love.Everyone resonated to these naive melodramas, and enjoyed the simple harmonies of the guitar or the banjo, and especially the harmonica (which Jim played like a virtuoso). Nobody missed the synthesizers and the high-tech music which they loved so much back in town.

When the meal was over, Grant, the mission chief stood up and put a beaker under the crushed-ice dispenser. An avalanche of ice cubes fell in, to which he added a shot of tequila. He raised his glass, " Well, troops, tomorrow is the big day!"

It was a long and uncomfortable night. The forest was restless—agitated in great oscillations. The tree trunks were swelling as if in pain. The waters grew darker. Treachery. Part of the shore was slowly collapsing under him, burying him. He woke up often—nightmares!

In the morning, they boarded the workboat and anchored at the mouth of the river, near a red marker: "No Fishing." Continuing in the skiff, they went a little upstream and tied up to a rock. A bald eagle was circling overhead.

With great care, they set a large rectangular box on the shore. They removed first the lid, then the sides. The great beast was resting on the box's polished floor. Its powerful four-foot-long (1.2 m) body was shiny under the layer of artificial mucous—the result of many months of work. Sunlight glinted in red and gold hues from the scales, exactly mimicking the natural appearance of the animal.

Under the scales, a strong and flexible plastic skin faithfully responded to the motions created by the complex internal machinery.

He was in the water to his midriff when they tilted the box. The animal's head was pointing towards him.

Grant called to Myriam, the bio-engineer, "Your move."

She was now holding the ichtyoid by the tail to keep it from slipping too quickly into the river. Grant worked the levers on the control box. The tail and the fins moved, the jaw opened. Lying first sideways in the water, the body twisted and took on a more normal attitude. Inside, relays opened, and cameras and sensors were activated, transmitting information to the data-loggers on the workboat.

Suddenly, the ichtyoid leapt forward, escaping its handlers, and almost immediately disappearing from sight. It was now up to Grant to follow on the screen the underwater adventure transmitted by the electronic eyes of the "Beast," as they called it.

A few days after this final test, they came back to the river mouth. The salmon runs had now entered the river in their thousands to spawn—and to die. But the Beast would not die. It would never die. It would also spread its genetically improved sperm on thousands of eggs from which would hatch bigger, stronger, more numerous salmon. That was the goal of the mission: to improve these valuable resources, in both quality and numbers.

The powerful ichtyoid spread its milt, mimicking even in its apparent fatigue the other salmon at the end of their forces. It then drifted downstream before rushing to fertilize another brood. But as the days passed, other fish began to avoid it. Soon it found itself completely shunned.

Sometimes, moved by some mysterious instinct, a real salmon would disperse or crush with its tail those eggs recently fertilized by the robot.

"This is how the species defends itself against the intruder," remarked Grant, observing the scene on his screen.

As to the diver, he witnessed the process to its very conclusion with an air of detachment which almost worried him. As the following summer came, he joined the team again.

There were soon noticeable changes. Some fish, most of them actually, were born with a head that grew to a disproportionate size, with big bulgy eyes. Others suffered from a narrowing of the fins or a shortening of the tail. The experiment appeared to have negative results. The Beast was pulled out of the water. Its slow and clumsy offspring was quickly eaten by seals, otters and sea lions.

They were locking up the shack. Grant gave his briefcase to Jim, the pilot. Standing on the float, Jim slowly scanned the edge of the forest. The bark seemed featureless. The quiet waters soothingly inviting. He pulled himself onboard. The plane took off effortlessly.

It was over. He felt empty. One could say that he was an accomplice. But he didn't feel guilty.

***

When one is tired of numbers and statistics, it is tempting to indulge in fiction. It frees the imagination without betraying the author's mission, which guides him throughout his work. The story offers a depth of perception which is lacking in a simple account of facts and figures.

And really, what could one expect from an artificial fish, raised in a reservoir, without the benefit of a competitive environment? Some will see it as a scientific attempt to repair the damage caused by modern greed—the price to pay for material comfort and economic development.

However, the efforts and sums invested on behalf of the salmon exceed anything that science alone could command. There exist indeed impressive achievements, especially in the realm of fish ladders, in France as well as in America.

Please understand that I have no wish to call in question the utility of such projects. They certainly assist the salmon in their spawning migrations. Improved stocks may be introduced for the benefit of the fisher-folk, especially the Natives. In the Pacific Northwest, the First Nation tribes rely on the physical presence of the salmon to maintain some link with their traditional lifestyle. Without it, a deep degeneration menaces Indian culture, especially at the symbolic level.

Undoubtedly, the voice of the Clallam Indians stating that the Elwah River is part of their ancestral territory is more eloquent than a long harangue. Similarly powerful, even today, is the First Salmon ceremony performed on the Tulalip Reserve. The first salmon caught is carried on a tray of cedar boughs. After the ceremonial feast, the bones of the fish left on the tray are taken back to the river.

The Last Rites of the First Salmon

Is the power of the mighty salmon still as mighty as ever in the Pacific Northwest? One may well ask this, after seeing the changes brought about by the taming of the Celilo Falls at The Dalles. Haunting old photographs remind us of the former glory of the site and of the beauty of the wild river and the people that it nourished. In the words of William Dietrich:

> The primeval Columbia simply looked different. Not only was it swift, but it was narrower and studded with towers and castles of basalt. Only a few of these outcrops escaped dynamiting or drowning by dam reservoirs. Broad bars of flood-washed boulders, gray sand beaches lined with cedar canoes, bright white water rapids where waves mounted higher than a man: all are gone.[5]

Nevertheless, even today the Indians continue to honor the salmon. In mid-spring, celebrations are held in the traditional longhouses, long wooden houses with a roof shaped like an inverted

---

5. W. Dietrich, *Northwest Passage*, p. 91.

"V." These differ from the feasts of old only in the fact that they also celebrate the gathering of edible roots.

By the way, I noticed that when my Dutch friends Tish and Pieter in Seattle, or Dave and Elizabeth on Vashon Island, bought a salmon, it was a rather solemn occasion; they carefully examined the fish for its color and texture and then ate it respectfully. Respect is clearly the correct word, which reflects the description given by the Methodist missionary Henry Perkins of a ritual he observed at Celilo Falls, shortly before 1843:

> the "tu-a-ti-ma" [twáti-ma, "Indian doctors"]—or medicine men—as they are sometimes called by the whites, practice a sort of invocatory ceremony on the first arrival of the salmon in the spring. Before any of the common people are permitted to boil, or even to cut the flesh of the salmon transversely for any purpose, the "tu-a-ti" [twáti]—medicine man of the village—assembles the people, and after invoking the "Tah" [taax] or the particular spirit which presides over the salmon, and who they suppose can make it a prosperous year or otherwise, takes a fish just caught, and wrings off its head. The blood, which flows from the fish, he catches in a basin, or small dish, and sets it aside. He then cuts the salmon transversely into small pieces, and boils. The way is thus opened for any one else to do the same. Joy and rejoicing circulate through the village, and the people now boil and eat to their heart's content.
>
> But I wish to call your attention to the blood. This is considered to be "aut-ni" [áwt-ni]—or as we would say sacred, or hallowed, or sanctified, i.e., it is sacredly set apart and carefully garded for five days, when it is carried out, waved in the direction in which they wish the fish to run, and then carefully poured into the water.[6]

There is no doubt that Henry Perkins was a careful and faithful observer, as all his reports confirm. He avoids commenting on what he describes and leaves us free to speculate. Did he as a Christian missionary draw a link between Celilo's sacred salmon and the

---

6. H. Perkins quoted by Eugene S. Hunn, Nch'i-Wána, *(The Big River)*, p. 153.

Fish-God of which Jesus, the fisherman of men, was the apostle? Did he merely describe a pagan rite aiming at ensuring the abundance of a resource which was the basis of their economy? Did he simply find childish a ceremony that others of his cloth regarded as either exotic or dangerously superstitious?

After all, even though the Indians were dying like flies from contagious diseases (dysentery, small pox) they were not, for all that, turning to Christianity. One of Perkins' colleagues complained that in seven years of apostolate he had not converted a single soul, a sign that in spite of superficial similarities, the Native rituals remained in their essence quite remote from the faith preached by the missionaries. It would then seem that Perkins had perceived the pagan character of the first salmon ceremony: the celebration of birth, life and rebirth at the very moment of the death of the Great Fish.

Through this seasonal ritual, the Natives of the Columbia participated in the annual cycle, itself nested in longer periodicities. Such a cyclic view of the world could only worry the nineteenth century propagators of the Christian faith, be they Methodists or Lutherans.

Some 100 years later, in 1957, the construction of the dam at The Dalles would spell the end of all such worries. It meant the elimination of the Celilo Falls, and with it the likely extinction of a certain tribal lifestyle. Finally, both the river and the Indians would be tamed by the skill of the engineers. At that time Tommy Thomson was hereditary salmon chief of the Celilo tribe. It was he who decided on the date of the opening of the fishery, and of the length of the fishing day as a function of run size.

When he heard about the project, Thomson was stupefied. He declared that he couldn't imagine how he could live if a dam submerged the fishing sites. It was the only food he could rely on, his only source of revenue, and he was there to protect it.

The engineers of course ignored Chief Thomson's objections. He presided for the last time at the First Salmon ritual on April 20, 1956. Until the very last day, the Indians fished from their rickety platforms rising above the foam of the dying falls. Nevertheless, 40 years later, the roar of the falls still resonates in the memory of the Natives. "If you listen, you can still hear its roar. If you inhale, the fragrances of mist and fish and water come back again."

Celilo Falls shortly before destruction by the Dalles Dam on March 10, 1957.

A view of the Columbia River gorge, upstream from the Dalles.

Chapter 19

The Apple Trees of the Columbia

Why bring up apple trees when I don't even like apples? All I really enjoy about them is their color. As I am writing this in the height of summer, I suppose that the images that inhabit my mind are influenced by the seasons.

It was indeed in the summer that I discovered the orchards of the Columbia, from Wenatchee to Okanogan and beyond, into Canada. The sun was beating on the valley with temperatures easily reaching 100°F (nearly 40°C). In the winter, under the snow, the leafless orchards were but a shadow of their summer glory.

To conclude our introduction to the Pacific Northwest, we travel up the Columbia to its midreach, between the Oregon border and the Colville Indian reservation, an area with which I have become quite familiar and which I recall with endless pleasure. I know that in this month of August, my friend Ed Fuchs frequently leaves his home in Riverside to exhibit and sell his gold prospecting equipment at fairs, trade shows and flea markets. Summer is orchard and gold-digging season.

However, as with fishing, the Indians were elbowed aside in this great dry garden where they used to carry on their nomadic existence. The shores of the Okanogan and the Columbia were domesticated with the rivers' waters. Enormous apple orchards were planted, enriching dynasties of apple merchants.

Apples are symbols of health and wealth: "An apple a day keeps the doctor away." New York City is the " Big Apple," ready to satisfy all appetites. Apple Computers link information technology with children's rosy cheeks. The link between health and wealth is at the heart of the fruit. In ancient Rome, Vertumnus—from the latin verb *vertere*: to change—was the god of fruit trees. To seduce one of the cousins of Pomona, the goddess of fruits, he took on various shapes: field worker, grape picker, harvester. He finally overcame her suspicion by masquerading as an old woman. He was also said to have diverted the course of the Tiber River. When shown in the pantheon of gods, he is depicted in ever changing form.

The time has come to tell the story of one the pioneers of the apple orchards, Hiram Francis Smith, born in 1829 in the state of Maine. Starting out as a printer's apprentice, he worked for the big dailies of New York and Detroit. Drawn to California by the gold rush of 1849, he went on to handle freight at The Dalles before joining the Cariboo gold rush of 1858. Unsuccessful in the Cariboo, he settled on the shore of Osoyoos Lake, bought land from a local Indian chief, Tonasket, and took an Indian wife, Mary. He first built a store, serving both the prospectors and the general needs of new settlers. He quickly diversified his operations, transporting goods by mule trains, raising cattle, and keeping a small herd of scruffy but hardy pack horses. After a trip to Hope, in British Columbia, he returned home with 1,200 vine and fruit tree cuttings.

Sternwheeler steamboat *Okanogan* on the Columbia River run between Wenatchee and the Canadian border (1913).

Smith was making his mark; he was soon to be nicknamed "Okanogan Smith." In the words of his friend, Frank Streamer:

> The whole of the Okanogan country is grand, picturesque and attractive. The mountain dells are filled with springs and lakes; grain and grasses of all kinds flourish. The river itself is very pretty, placid and primitive. Its banks, on both sides being embroidered with a narrow strip of willows, cottonwoods, hawthorne, wild cherries and underbrush, being very difficult to pass through and it is only in one place where one can get water by having a clear space to sloping edge at the river. The water is dark, discoloured and impregnated with alkali; not at all wholesome, but the mountain streams are all cool, clear and very invigorating.

The entire country taken as a whole, from the mouth of the river at the Columbia to the Sooyoos Lake, a distance of 90 miles [145 km], may be set down as the best in Washington for grazing and gardening but it is no grain country. All the mountain uplands for many miles in width are well dotted with lakes, beautiful dells, plenty of aspens and other tree growths. This, then, is the Indian Moses' great reservation, superior in grandeur and worth, to the Jewish Moses' happy land of Canaan.

On the side of Highway 97, between Oroville and the Canadian border, there is a historical panel describing Smith's life which essentially reads as follows:

Okanogan Smith

Hiram F. (Okanogan) Smith settled on the opposite shore of Lake Osoyoos at the end of the 1850s, becoming one of the first white colonists of the county. For many years, his trading post, the only one of the area, was a welcome sign for traders and traveling miners. In 1857, Smith planted 1,200 small apple trees which he had brought back from Hope, British Columbia. These trees were at the origin of the apple growing industry in Washington State. Eleven of them continue to bear fruit. Miner, pioneer of the cattle industry, Smith occupied important posts, first in the Territorial, then in the State Legislature until his death in 1893.

Succinct as they are, these historical panels are extremely informative. In the United States, history is invisible: there are no ruins. Except in rare cases, the old wooden forts, churches and city halls have disappeared. For example, at the confluence of the Okanogan River and the Columbia, there stood Fort Okanogan—an Indian word for " meeting place." The fort was erected in a strategic position, dominating a broad panorama overlooking the joined streams. New York fur trader Jacob Astor built a trading post there in 1810. This post was bought up by the Montreal-based Northwest Company in 1813 and became property of the Hudson Bay Company when it merged with the former in 1821. Fort Okanogan was an important commercial establishment. It attracted fur traders and prospectors; the Indians bought cattle and sold horses there; it served as a base for catholic missionary Father DeSmet. Today, the fort is just a dot on the map.

In the same vein, what's left today of Hiram F. Smith? A few hundred-year-old trees and a crumbling old barn—an eyesore to some. Nevertheless, he is well remembered. Americans readily engage in exaggerated praise: regional historian Sandra Hildebrand speaks of Smith, in many more words, as one of the immortals in the pantheon of the pioneers of agriculture.

Hiram Smith undoubtedly deserved all the praise he has received. His friends all spoke of his human and moral virtues. Surrounded by Indians—whom he always respected—and rough adventurers, he never lost his civility, his strict honesty and his adaptability. He sought privacy, preferring to work alone or to reflect quietly, without being disturbed, except by his close friends. One of them, Francis Streamer, has left notes warmly describing life in the patriarch's compound.

Mary, his Indian wife, their two daughters and Indian in-laws living on the property were always hard at work. There were rarely fewer than 20 people at the Smiths' table; no one left the table hungry. The word of the patriarch was respected by all, whites and Indians. His trading post was an oasis for all travelers, be they traders, mule-train leaders or trappers.

Sitting under an apple tree Francis Streamer wrote in his notebook:

> I have had extraordinary kindness and courtesies extended to me since I came here. I have enjoyed the fruits of the orchard, there being plenty of ripe apples of extra size and flavor, and plenty of corn and potatoes. Mary, the lady of the home, made me a very neat pair of moccasins for my office work which I wear now in great comfort. I have bathed and washed all my clothes, even hat and suspenders in the pure fresh waters and am today a renewed and invigorated man, happy as a lark and as contented as a humming bird…[1]

---

1. Quoted by Sandra Hildebrand, *Treasure in the Okanagan*, pp. 111 and 112.

Pioneer orchardist Hiram Smith brings to mind another famous man: John Chapman (1774–1845), better known as Johnny Appleseed. An orphan, raised in Massachusetts, Johnny drifted west with the frontier, planting orchards as far as Ohio, helping many pioneers. Unmarried, a vegetarian, he was also a friend of the Indians and of animals. He became a hero of American folklore, a deep fountain of literary inspiration.

In his famous "Martian Chronicles," Ray Bradbury honors Chapman's memory through a character who plants thousand of trees on the red planet.

> There were so many things a tree could do: add color, provide shade, drop fruit, or become a children's playground, a whole sky universe to climb and hang from; an architecture of food and pleasure, that was a tree.[2]

and further:

> In school they told a story about Johnny Appleseed walking across America planting apple trees. Well, I'm doing more. I'm planting oaks, elms, and maples, every kind of tree, aspens and deodars and chestnuts. Instead of making just fruit for the stomach, I'm making air for the lungs. When those trees grow up some year, think of the oxygen they'll make![3]

In his role as a hero of colonization, Chapman figures as a kind of holy man, an American archetype. However, his aura of holiness is linked to a peculiar characteristic: his spiritualist teachings. Before continuing on his way, Chapman left the settlers with a handful of seeds and also brochures exposing the ideas of Swedenborg.

Emmanuel Swedenborg (1688–1722), the son of a Swedish pastor, was a priest, organist, mathematician, physicist and visionary whose influence in Europe was widespread, as seen in the works of Balzac, Beaudelaire, Nerval, George Sand, Kant, Strindberg, Novalis,

---

2. Ray Bradbury, *The Martian Chronicles*, p. 73.
3. Ray Bradbury, *The Martian Chronicles*, p. 74.

Paul Valéry and many others. His works were to become even more popular in the newly settled lands of America, particularly because they did not require a break with traditional religion, but suggested concepts likely to appeal to the New World's pioneers.

Swedenborg's ideas are based on Christian doctrine, but diverge from the catholic or protestant mainstream. In the words of Marie-Madeleine Davy, a specialist in mystical theology, "Christ reestablished the balance between heaven and hell, but since his coming, additional misdeeds have been perpetrated and a new Judgment will be necessary."

U.S. Johnny Appleseed commemorative stamp. (Issued in 1966.)

Swendenborg's originality resided in his exploration of the parallel worlds where he ventured following his dreams and visions. He describes these worlds with the precision of a geographer of the celestial spheres, claiming that he has been chosen as an intermediary to explain to mankind the meaning of God's words. He links the macrocosm with the microcosm. In M.M. Davy's words, "the created universe contains in itself the divine image," so that the material world becomes a representation of the spiritual realm. But man has focused more and more on the tangible world, gradually becoming blind to the world of the spirit.

In America, the Church of the New Jerusalem spread the ideas of the Swedish visionary. In the Pacific Northwest, an area favored by nordic emigrants because of climatic similarities to their homeland, there was great interest in these new views. The settlers, having freed themselves from the rigid framework of traditional order, were building a new world of splendid wilderness. Swedenborg was telling them that spiritual causes affected material events, that "the Lord appeared to each one in a manner adapted to their own individual capacity," and that the idea of change was the active essence of the world.

Johnny Appleseed, the traveling gardener, was, to paraphrase Swedenborg, the very prototype of how a man should be, living with his spirit engaged in the spiritual and his body in the material world. This kind of attitude, of dynamic spirituality and unorthodox idealism emerges today in the behaviour of many Americans. It is evident in the person of Hiram Smith, whom we know today thanks to the writing of his friend and great traveler, Francis Marion Streamer.

Chapter 20
The Lonely Pedestrian

Francis Marion Streamer was born in 1834 in New Hampshire. He was a man of many trades. He first learned that of a harness-maker, then became manager of a store, and enrolled in the army of the North at the beginning of the Civil War, in 1861. After the war, he worked in various states, from New York to Iowa, as a scout for the army, a teacher to the Chippewa Indians, a farmhand, and an insurance salesman. He then became a journalist in Omaha, Nebraska, and later in Sherman, Wyoming.

One day, at the summit of the Black Hills (3,300 ft [700 m]) near Sherman, he heard, far above in the sky, "the prayers of the coronation of the sun." From then on, the "heliocentric principles" would guide his life.

On October 11, 1875, after leaving his gold-pommeled cane to the care of his friends in Pennsylvania, he took his leave and started on foot for the Washington Territory. A year later, at the age of 42 he arrived at The Dalles. His notebooks, relating his travel adventures, have been a treasure trove for historians. He had to face, cold, heat, hunger and thirst.

Wednesday Evening June 19th, 1877: I famished, fainted and fell by the trailside this whole day. Ate wet mud and dug a hole in an alkaline pond, got some water, then came on to this old deserted Indian camp. Arrived at night and found a clean water creek, made a bed on dry tule mats the Indians had left and slept chilly.

Friday 3PM June 21, 1877: I cannot account for the strange travel. Something leads me, sustains me and cares for me.[1]

---

1. Ann Briley, *Lonely Pedestrian: Francis Marion Streamer*, p. 46–47.

Whenever doubt sets in, he receives an Order, a Solar Direction which he sometimes calls Elijah's Decree. He settles temporarily in Ellensburg where he spends six months as a school teacher. During those months, he also writes articles for national newspapers eager for sensational news from the Far West. However, Streamer takes the side of the Indians, an unpopular attitude at the time.

One of his notebooks, entitled *The Book of Wisdom,* describes in minute detail the life of the Indians, in their camps, salmon fishing and berry gathering. The author describes how an Indian chief transforms his wigwam into a sweat lodge:

> I watched him heat the stones, make a hole for them, floor his wigwam with fir leaves *(sic)*, cover the bake oven with tent canvas and blankets, then get into the cold spring water, bathe and then go into the hot house which was so hot I could hardly hold my hand in it, and then sweat himself as did others of the party, men and women...[2]

In August 1880, Streamer headed toward the north of the territory and arrives in the Okanogan Valley—eight years before the creation of the county of the same name. A very few white men had settled in the land, prospecting or raising cattle. Most of them had taken Indian wives and were known as "squaw men," a term which would soon become pejorative, in spite of Streamer's admiration for the Indians and his efforts to share these feelings with other newcomers.

> This generation and age is far inferior to that generation of giants in form and intellect that inhabited this country centuries ago, and who had more knowledge of the electric currents and their uses than a Franklin, a Morse or an Edison ever dreamed of.[3]

In 1883, he hung his bag at his old friend Okanogan Smith's ranch and stayed for good. He became tutor to four little Indian

---

2. Ann Briley, loc. cit, p 58.
3. ibid. p. 79.

girls, as well as secretary in charge of all commercial correspondence and, through his notebooks, unofficial biographer. A man of many skills, as we have already noticed, he was also a licensed notary.

Not only was Streamer outstanding in the performance of his many duties, he also staunchly defended the rights of his Indian friends. He appealed to his wide circle of friends to defend Chief Moses against the cravings of the big timber companies, eyeing the lands belonging to his people's reserve.

Occasionally, Streamer went on a binge, a fact which he freely admits. Historian Ann Briley wonders:

> But was it totally a force for evil? In our present theories of mental therapy, it might be construed that this recourse to alcohol was a self-administered tranquilizer which saved him from more violent forms of mania. Were he living today, we might load him with drugs, which would render him harmless and dull, but he would never have achieved the vivid life he led, which left his strange chaotic contribution to history.[4]

I take pleasure in imagining an encounter with Arthur Rimbaud (1854–1891), the foot-loose poet, versifying after a few drinks with his friend Frank,

> *Sweet as the Lord of the cedars and the hyssop,*
> *I piss towards the sepia heaven, far above, far away,*
> *Confident in the permission of the tall heliotropes.*[5]

Nevertheless, Streamer's circle of friends in the Pacific Northwest remained faithful to his last days. He had become a familiar character, with his bag of books over his shoulder. Spending only a few days here and there, wandering from his home base at the command of the Sun. His bag, a subject of wonder and respect, contained his notebooks as well as other manuscripts which he always

---

4. Ann Briley, loc. cit., p 105.
5. Arthur Rimbaud's poem: "Oraison du Soir." Translation, P. LeBlond.

carried with him, his "scrolls," with titles like *The Book of Revelations*, *The Scroll of Oracles and Voices from the Air*, and *The Book of Wisdom*. He left some of these with a friend in Omaha who submitted them to the *World Herald* for publication. The newspaper editor sent them back with the comment that they were, "too comprehensive and too enlarging upon too many subjects."

Perhaps a contributing factor to this total dismissal of Streamer's writings might have been his exposure of the double-talk of the bureaucrats of his day. When they spoke, for example, of "efforts to concentrate the dispersed Indians into population centers," what they really meant was that the nomadic people of Chief Moses would be tamed and packed into a settlement along the Nespelem River (today, the heart of the Colville Indian reservation).

Visiting some old friends in Omaha in 1890, Streamer remarked, without much surprise:

> the Puritanical fathers had got possession of the mayorality of the city; that Quakers were driven out, Ponca Indians carried away to Indian Territory to die with malaria...
>
> The entire trade of the city had relapsed into the dyspeptic condition of Women's Temperance Unions, Salvation Armies...[6]

In 1893, good old Samaritan Okanogan Smith died. In the months that followed, Streamer did his best to ensure that his legal heirs would not be despoiled of their inheritance. He continued wandering from friend to friend, filling his notebooks with poems and metaphysical questions.

> Does a spirit inhabit a body? No! Spirit is Ether in air and controls tone of voice only. Soul is love in Sun-ray and attracts the eye upward. Kindness is a magnetic impression by inhalation of Sun warmth on forehead, and Pity is a gift from the personal Solarity to the earthly being of clean speech, clear tone and tearful eye.[7]

---

6. ibid. p. 113.
7. Ann Briley, loc. cit. p. 154.

Strange words one might think! Hardly so, if one recalls the spiritualist climate fostered as the time by Swedenborg's New Jerusalem church, among others. Another passage illustrates some of the views that puzzled his contemporaries. Streamer expresses his opinion of the contents of the newspapers sent to him by his sister in Iowa:

> it was kind of you to give me the opportunity to thus interview Iowa ideas of millennial transformations, men and women's peculiar views of this electric era. Of course, Mary—two thousand years ago in old Mesopotamia I taught the same theology.[8]

F.M. Streamer, chronicler, physician, school teacher, lonely Don Quixote, poet and philosopher, spent the last 14 years of his life in the old-folks home of Medical Lake (in Spokane, Washington). He died there from tuberculosis on January 18, 1912. It was said of him that he spent his life as a visionary recording his own visions.

I am surprised that Frank Streamer's path never crossed that of a real live sasquatch. He certainly knew that giants had lived in this land eons ago.

---

8. Ann Briley, loc. cit. p. 155.

Chapter 21

A Visit to the Museum of the Colville Confederated Tribes

It's a small museum, very neatly organized. One can see traditional garments of the Okanogan tribes, jewelry, basketry, a diorama of the natural habitat, and black-and-white pictures of five generations of Indians. The museum is in the village of Coulee Dam; the dam itself is a few hundred yards away.

The director, Andy Joseph, was expecting me. He is an Okanogan Indian, author of a teaching method—a book and an audio-cassette—of his native language, Interior Salish. Linguists and ethnologists regularly visit him and spend time in Coulee Dam to take advantage of his expertise. Andy is about 60 years old; he wears jeans and a denim jacket, moves quietly, seemingly without bending his legs, but surprisingly quickly. His deeply wrinkled face, flanked by a pair of long salt-and-pepper tresses, is brightened by a whimsical smile.

Andy is expecting me because I had made an appointment. His Indian friends had mentioned me; he knows what brings me. He leads me to his office, in the basement, where two young Indian women are working. One of them, slightly plumpish with long wavy black hair gives me the most gorgeous smile; the other, with her fine features and slightly upturned nose is equally beautiful. However, over the next two hours, sitting across from Andy, I keep my eyes on him alone so as not to lose a word of what he tells me. It is difficult to escape his gaze and I need to listen attentively as he speaks in a very low monotone. There is a joke about him: "With Andy, you can't turn up the volume!"

Andy was obviously going somewhere with his narration, but I wondered where. His talk was a rather enigmatic monologue, inviting few questions. I quote from my journal:

Wednesday June 30, 1994

First, Andy speaks of his childhood, and his vision-quest for

Andy Joseph, Sr.

his guardian spirit. Although it is an animal spirit, it will manifest itself in human form. This is why Chief Seattle recommended to pay as much respect to animals as to people. Unfortunately, development has driven the animals further and further away.

One should know that in certain tribes (the Wenatchee, the Moses) the bear is a brother. Among the Okanogan (Andy's tribe), bear meat is not to be wasted. Andy had killed many of them. His brother-in-law had served bear meat at a funeral. The guests thought that it was venison. Unfortunately, one of Andy's nephews fidgeted on his chair, fell back and killed his younger brother on the spot.

Andy said, "I could have prevented that and I didn't do it. Why?" Was it because the sight of a bear (the brother? the double?) may be a harbinger of death?

The vision quest is a hardening experience for a seven- to eight

-year-old child, made to walk to the end of the compound, at the edge of the forest, to check that the gate is properly closed. Yes it is! But why this long and scary trek at night? The child does not understand, until the day... when the guardian spirit manifests itself. [at that point I completed the unfinished sentence myself]. The Indians are sight-trained to recognize their vision (in Salish: *week sst*). They are prepared for encountering their animal spirit.

Having learned that the Steller's jay occupies a special place in the mind of the Interior Salish people, I asked Andy about it. According to him, the jay can find anyone and anything that is lost. It is one of those spirits with great powers. Andy relates the story of a small "jay-woman"—very well turned out, by the way—who was capable of remarkable achievements. One day, she climbed up onto a roof without means of access, such as a trap door, a ladder or a climbing post. She refused to explain how she had done it.

Andy now returns to the main topic. As a child, he saw the *Swenatum* (the Interior Salish word for sasquatch/bigfoot). He was in the company of his adoptive mother, who was blind. A white man was driving the car. He asked, in Salish,

"Why is it here?"

His mother answered, "He offers his presence to you."

Swenatum manifested itself frequently to Andy, as it does to the Indians generally. He will never forget the sasquatch song that he was taught, a song without words, to be hummed.

I asked him, "When do you sing it?" "Whenever necessary," he answered.

Andy's voice was but a whisper. His speech, peppered with Salish words, sometimes lacked continuity. I listened attentively.

We are an old people, often despoiled. We are now trying to return to the graves from which they were taken some historical artifacts of great historical value. We do not systematically collect objects. For example, we do not pick up old arrowheads. If they are on the ground, it is because they have missed their goal. One must let the survivor be and not reuse the arrowheads.

Colville Confederated Tribes insignia. (Photo by Donald Healy)

Do you know who invented the arrowhead? It was Coyote, one of the most disagreeable beings. He is the one who scatters its droppings and even can even die from eating them...being revived each time by Fox. Coyote is supremely noxious, especially since he has a strong creative power [like the Devil today; thus people say, "Coyote made me do this, or that"].

One day, two brothers, always disputing, came to blows. Coyote took hold of two gigantic rocks—these enormous basaltic masses seen along the Columbia. He banged them together with such noise that the brothers ended their fight. Many fragments of rock fell to the ground. Coyote picked them up; he had just invented arrowheads.

Sin Ku Leep, the Coyote, is an inventor, for sure, but sometimes his skills turn against him. As on the day when he dressed himself up in deer skins to seduce a local beauty. Her four brothers went after him into the mountains. They finally found him stuffing himself on prairie dogs. They caught him and threw him in the water. He fell to the bottom of the lake, but Coyote had the power to transform himself. All he had to do was to say the word "fish" or "turtle" to become one. Alas! His tongue slipped and he said "rock" and sank to the bottom.

An Indian knows that he has to hone his sight to see the way. He engages in an ancient practice, the sweat-lodge, to purify body and spirit. "We are a very ancient people," says Andy. "Like Swenatum, we were here 12,000 years ago." Andy stressed:

> I have seen Swenatum many times, and in the daytime. The last time, a few years ago, we were at the edge of the forest, not far from here, preparing to leave in my pick-up truck. My three sons, aged eight, 12 and 17 years, were with me. On the left, there was a wide-open sloping area; I saw a dark silhouette profiled against the sky. It was not a bear. I started the truck, driving uphill slowly to approach it, without the slightest fear. My youngest son didn't seem to care; the middle one was afraid and wanted to go home. Their older brother, standing behind in the box, his eyes riveted on the creature, was urging me to move faster. But after a few moments, Swenatum disappeared. It was my older son who was most deeply touched by the encounter with Swenatum. What impact will this event have on his spiritual development?

The museum was now closing. Andy continued his story at a table in a nearby café, touching on many other aspects of his life, confidences for which I remain grateful.

I soon had to leave, being expected in the nearby Indian village of Nespelem. I had already experienced a number of encounters. Others would come, often rich in information. There would be more fables, other lessons.

Some final words before leaving the marvels, dramas and simple splendors of the Columbia, the kingdom of Swenatum.

People are irresistibly affected by the scenery that surrounds them. For many, although they may become acclimatized, the surroundings never becomes commonplace. Acclimatization helps the newcomer get a feel for the scale of the country and become comfortable with its dimensions. Fascination leads to paralysis, while some familiarity with the land frees the spirit, brings about a serene admiration, and opens the door to further learning.

The ground bears the marks of long and perplexing cycles of glaciations and volcanic eruptions. Curiously, one seems to apprehend almost spontaneously these gigantic time periods—in my case, I have related how Ed Fuchs helped me understand them. The shorter cycles, of mere tens of millions of years, are easier to grasp, partly because of the clearer traces they have left (fossils, erosion, bones) and because of the continuity of their bygone fauna and flora with those of our times.

My own travels have been quite modest. I sometimes have the impression that I remained at rest, while the Earth changed under my eyes. The Pacific Northwest is such a special place. Without much of a shift in my point of observation, I witnessed the alternation of fire and ice, the ebb and flow of the waters, the growth of the forests, the waxing and waning of populations, and the wandering of the tribes, of which many—or rather their descendants—still exist today.

The past, even the most remote past, and its tracks—like those of the sasquatch—leave such a deep impression on the present day that it is impossible not to recall them, even when describing the events and affairs of today's cycle, that of modern man.

Within this space-timeframe there live—or lived until recently—surprising characters, either born in the area, or shaped by it. Learning about their lives enriches our understanding of the region. Through Okanogan Smith, for example, we can witness the transformation of some of the countryside.

People of old witnessed the Columbia River in its wild, untamed state; they were sometimes its beneficiaries, sometimes its victims. They then decided to alter its course, a hazard of the march of civilization which significantly impacts our enquiry. Once tamed, the Columbia brought great blessings to some, but destroyed the way of life of others, especially the People of the River. Fishing was their means of subsistence. The river provided the décor, determined the rhythms of their life and nourished their spirit. It was a part of the soul of the People of the River, who lived in close harmony with nature. I have often thought that the sasquatch was subject to the same constraints: the dams on the Columbia led to the disappearance of a vital resource, the salmon. Those who most closely depend on it: the bears, the Indians, the sasquatch, and in some way also the poets and the dreamers, are being forced into exile and may even disappear.

This, then, as it stands at the end of the twentieth century, is the realm of bigfoot, born three million years ago and a contemporary, during the Pliocene, of *Homo habilis* and *Homo erectus*. According to anthropologist Grover Krantz, bigfoot would be a kind of *Giganthopithecus,* a giant ape-man, who survived through the Pleistocene in the company of "ancient" *Homo sapiens, Homo sapiens Neanderthalensis* and finally *Homo sapiens sapiens,* yours truly.

Part III

Wild Men and Men of Science

The crystal reveals all, from stone to plant
or coral, from the vegetal to the animal,
otherwise the same leap, from animal to man.
To distinguish man from beast,
I am told that I alone stand erect: a false antinomy.
Both the bear and the kangaroo also stand,
So do monkeys and snakes...armless as they are.
Meanwhile, baby, cat, dog, weasel,
Beaver and bird soon drop what they caught.
It's then Word, the abstract, that makes me sacred,
But from the amoeba to the elephant, some guard
Warns of danger, of food—none for now!
And their people follow them
While mankind damns its prophet. That very threshold
The plant dared cross, the stone respected.

JEAN-CHARLES PICHON
L'Animal et l'homme [*The Animal and Man*]
Les Litanies des Dieux morts (2000)
[*The Litanies of Bygone Gods*]

Chapter 22

The Sasquatch: A Few Words Before Hearing from the Scientists

Sasquatch investigations must rely not only on eyewitness reports but also on consulting a great number of newspapers, novels, and scientific studies. No stone should be left unturned. The most innocuous or even the most unappealing publication may show the way, stimulate thinking and sharpen the researcher's curiosity.

One day, in a book store, I discovered three small books entitled *The Sasquatch Have Their Ways,* a collection of stories published privately by the author/editor, Steve Heinzen, in Everett, Washington. The cover of the third volume boldly proclaimed: "Bigfoot! He's still out there!" Half that volume was an account by reporter Bonnie West of her enquiry on the presence of bigfoot in Grays Harbor County.[1] Bonnie wrote:

> Whether one believes or not in the existence of the sasquatch, the fact remains that among the 6,500 inhabitants of Grays Harbor County, some number have seen, or heard, or smelled something very unusual. Their experiences raise questions. Is the sasquatch strictly a folktale, or is it some zoological oddity? And if one were to actually find one of these bipeds, dead or alive, would it be classified as closest to the species which takes pride in its beauty queens, or would it be placed with the primates, like those Jane Goodall has been studying for some years in Africa?

Speaking to Steve Heinzen on the phone, he suggested I meet Bonnie West, which I did a few days later, in Elma, Washington. She had gathered together a few sasquatch investigators. One of them

---

1. Bonnie West's interview with the author in *The North Coast News,* July 26, 1995 is reproduced in Appendix 1.

was Fred Bradshaw, former adjunct to the sheriff, who related how, in the days when he was still on the job, in 1987, he had heard the story of a man and his son who were pelted with basketball-sized rocks while canoeing on the North River.

Fred had decided to investigate and to camp near the site of the incident. He was accompanied by his son, Rick, and his sister-in-law Carol Howell. Here's what Fred stated:

> Rick and I had been trout fishing from five to eight o'clock. The dog was constantly whining. We couldn't get him to shut up. We should have known something was going on. We returned to our tents.
>
> Carol thought there was something strange around the camp. She could hear pebbles rolling down a steep slope as if an animal was climbing it. We went to have a look. It was a steep riverside bluff about 800 feet (250 m) high. Back at the camp, we sat around the fire. It was about eight-thirty and we could tell there was something lurking nearby. We then heard the cries of song birds, titmice among others, normally never heard at night. Twigs were creaking. There were muffled noises…
>
> We were startled by a loud breaking sound and a very large branch, about thirty feet (10 m) long crashed to the ground about 20 feet (6 m) from us. There was a rustling of leaves. Rick grabbed his flashlight, aimed it at the forest, and shouted, "Oh My God! Dad, Look at that!"
>
> Carol leaped up as I rummaged through the tent for my shotgun and my revolver.
>
> I could tell we were not welcome in the area! The creature in the beam of light was some 300 feet (approx. 100 m) away. Its head was hidden by a tree trunk, but they could see the rest of its body, down to the knees.
>
> The creature was nearly eight feet (250 cm) tall and weighed from six to seven hundred pounds (270–300 kg); it was light brown in color. It chest was very thick…My finger froze on the trigger. [Fred estimated that the arms were as much as 18 inches (45 cm) across and the thighs 20 to 24 inches (50–60 cm)].
>
> It had taken us three quarters of an hour to put up the

tents. It took us less than seven minutes to pack everything into the car and flee. Carol didn't want to leave. She was more curious than afraid.

We came back the next day to look the place over. The creature had approached within 130 feet (40 m) of us. We found some of its footprints. One of them was 18 inches (45 cm) long.

I took good notes of Fred's account. I became good friends with Fred, Bonnie and Carol and had the pleasure to visit them many times in Elma. Since his meeting with the sasquatch, Fred has become more sensitive to nature, and more respectful of the beings which it shelters. There is no question that he would ever shoot a sasquatch. It is also from that time that his relations with the neighboring Indian tribe, the Quinault, improved. So much so that Fred had been adopted as a member of the tribe, following some instruction, and has begun a secret initiation, similar to that of the "mystery covens" of antiquity. Fred gave some discreet hints—but that's another story.

Fred and Carol drove me to the spot where their encounter with the sasquatch had taken place, near a swift river. They showed me the tree which had masked the creature's head, validating in some way the veracity of their story. I had already noticed from many of the witnesses I had talked to, that specifying the exact location of an encounter seemed to provide a guarantee of the event's authenticity.

During my first meeting with the sasquatch hunters of Elma, I had met, over dinner conversation, a young man who had told me he had seen the sasquatch twice. Later that night, he took me in his car along a narrow forest road. He stopped to show me, in the glow of the headlights, the very tree that he used as a reference point for his first encounter. He had been 11 years old at the time. A year later, he again saw a sasquatch, in a small clearing on the side of a nearly impassable dirt road. He told me that he would have liked to build a log house on that spot. The area obviously exerted a strong attraction on him. He also told me that his mother had always refused to lend any credence to his stories, which pained him considerably.

Standing there, in the quiet of the night, in the middle of the forest, I found myself rather astonished, and somewhat more watchful. However, at no time did I have the impression that I was being the butt of a joke. Fred had also reassured me. "Trust him," he had said.

Examining the sites of encounters and listening to the witnesses merely sparks one's curiosity. I was seizing every opportunity to learn, familiarizing myself with the flora and fauna of the Pacific Northwest from Wendy Davis' children's book *Douglas Fir.* Another children's book, that I found most charming and nicely illustrated was P.R. Walker's *Bigfoot and Other Legendary Creatures.* Popular novels, such as Rainy Knight's *Critter* shed more light on rural social mores than on the creature itself, although it is interesting to note how the quest for the sasquatch follows in the footsteps of King Kong in public fantasy.

As yet another way to approach the study of the sasquatch, I delved into the world of those apes closest to people, the chimpanzees, companions of Jane Goodall for about 30 years in Tanzania. In words and photos, Jane Goodall describes the ape families she got to know, giving each individual member a name. I have often been fascinated by these creatures, their life and their moods. They were captivating by themselves, as apes, in the forests of East Africa where I once ventured; they also reminded me of human society, as Jean-Charles Pichon saw it:

> Every social grouping vaguely resembles an animal grouping. Some mimic a tribe of apes; others a warring pride of lions.[2]

Three images in particular attracted my attention; they feature a five-month-old chimpanzee, Galahad, already endowed with a strong personality. Jane Goodall expects him to grow up into the dominant male of his community. There he is, chewing on a stick. At first, he seems pensive, perhaps somewhat sad. But then, he takes

---

2. Jean-Charles Pichon (1920–2006), *Les Dieux Etrangers,* p. 38.
Pichon was a novelist, French philosopher and metaphysicist, specialist of myth and religion. For an overview of his life see:
<http://www.jeancharlespichon.com/npds/sections.php?op=viewarticle&artid=49)>

on an air of solid determination. As the sun sets through the trees on the shore of Lake Tanganyika, Jane Goodall reflects on Galahad:

> Tonight as I lie listening to the lapping of the lake on the shore, my mind will be filled with pictures of little Galahad.[3]

Galahad, the seeker of the Grail. Once more, the words of Jean-Charles Pichon come to mind:

> These innumerable species of soils, plants, animals, people; all these forms, colors...a voracity mixed with mercy, each suited to its own world, its own order, race, individual, which the puzzled scientist labels "natural equilibrium." This freedom of song, sprint, play, and work proper to larks, bisons, rabbits and beavers; or work, play, quest, song and freedom in each man and in each man differently...[4]

The North American Indians, having lost their freedom, and with it their work, play, song and dance, are recovering their independence. They return to a quest which they had never really abandoned. Among the Quinault—and many others—sasquatch is half man, half animal. Each individual is free to accept it either as myth or reality, a freedom of choice which reflects "the Indian way."

For the Quinault tribe, sasquatch is a protective being and a source of inspiration. The Quinault Nation Dancers honor him in their dances. Children especially love to play him. One of them, dressed in a rabbit-skin cape and wearing a sasquatch mask enters, hidden in a ring of other dancers. When they raise their arms, the other dancers notice the presence of the sasquatch and are struck motionless. Then, sasquatch dancer rushes them and teases them, as they believe the real being would.

I dreamed of taking on the part of the sasquatch dancer. What a role! That of a strong giant, smart, friendly, king of the mountain and the forest. But at the risk of looking like a buffoon, I would have

---

3. Jane Goodall, *The Chimpanzee Family Book*.
4. J.C. Pichon, *Les Dieux Etrangers*, p. 98.

to know its anatomy, its metabolism, its gait, its food, and even its ancestry. To embody a living creature of such stature, the mythical dimension would be poorly served by a mere caricature.

What do we know then of the sasquatch of flesh and bones? Who were its ancestors? Now is the time to gradually leave the floor to the "men of science."

Jane Goodall with Russian hominologist Dmitri Bayanov, Moscow, 1999. Jane has expressed a belief in the existence of sasquatch.

Chapter 23

Sasquatch and Science

Grover Krantz, whom we met before, a professor at Washington State University in Pullman, began to be interested in bigfoot reports in 1963. As an academic, he risked personal ridicule as well as endangering his career for his interest in a subject that many of his peers considered frivolous. Widely respected in his field—physical anthropology—as well as for his skills as a teacher, Grover stood his ground firmly in the face of many verbal and written attacks. Some accused him of having overstepped the limits of science. His courage and honesty earned him the respect of the great majority of bigfoot fans.

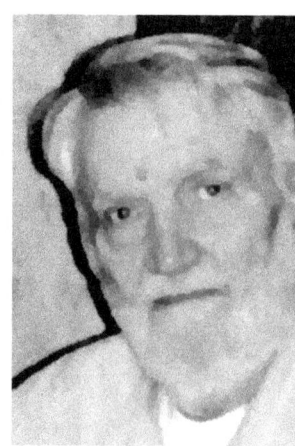

Dr. Grover Krantz

Tall, slightly bent over, bearded, of somber or at least preoccupied mien, Grover rarely went unnoticed. Dressed casually, in blue jeans and a faded khaki shirt, he walked like a man with a purpose. A purpose indeed, such as to endow him with an assurance approaching brutality in expressing his arguments. Speaking in a deep, nearly monotonous voice, Grover did not waste words; everything about him proclaimed that he was not joking. But he still had a sense of humor, or at least of irony.

Krantz's work *Big Footprints, A Scientific Enquiry into the Reality of Sasquatch* was written as an accurate description of knowledge accumulated on the sasquatch. The book is written in a style accessible to a broad public. If a few passages are rather dry, one can't blame the author for providing numerical dimensions and measures for the sake of accuracy. The analysis of the Patterson/

Gimlin film is, of course, the occasion for numerous diagrams and sketches.

In the introduction to his book, Krantz remarks that whenever one is asked about the existence of the sasquatch, the answer usually comes out either strongly positive or negative, rarely with any nuance, in spite of the poor knowledge of anthropology of the majority of the respondents. Such categorical responses reflect deeply rooted emotional reactions. For many people, including scientists, the human condition is seen in one of two perspectives:

> either we are simply an ordinary animal, one kind among many, or else we are a very unique form of life. While both observations are true to some degree, in many people's minds one view or the other is paramount.[1]

Those who believe that humans are "ordinary" animals think that people are the result of a rapid evolution from a relatively recent ancestor. On the other hand, those who believe that we are a "unique" animal think that human evolution has taken place over a very long time from very remote animal ancestors. A temporal chasm would therefore separate us from the animalhood from which we departed long, long ago. From that point of view, there is no place for an intermediate being between the animal and the human.

For the holders of the "ordinary animal" school of thought, it is quite plausible that there might exist an intermediate being, a further link joining us with the rest of nature. This is the missing link that would connect us to the animal world.

In Grover Krantz's view, both of these perspectives are erroneous. For him, the sasquatch is in no way an intermediary; in terms of its mental capacity and behavior, it is clearly an animal species. However, we have to remember that our knowledge of its behavior remains rather slim.

Three characteristics distinguish man from the rest of the animal kingdom: language, social organization, and the use of tools. Sasquatch pile up rocks, break or pull out bushes. It uses its hands

---

1. Krantz, *Big Footprints*, p. 11.

for fishing, gathering and hunting. It constructs beds and shelters of leaves and broken branches in the same way that the great apes construct their nests in trees. Little more is known since, in contrast to apes, there has never been a chance to observe the sasquatch's activities at leisure.

Dr. Grover Krantz's original book (1992), above, and right the revised and appended edition (1999).

Chapter 24

A Parenthesis

Chimpanzees use stones to break nuts, twigs to dig out termites. At Camp Leakey, in Borneo, Dr Biruté Galdika observed the enthusiasm with which a female orangutan was soaping and rinsing clothes; she had learned these skills simply by imitating people. The first tools used by people are dated at a million years ago, crafted by *Australopithecus*. Apes do not create stone tools in their natural habitat; they are however likely to succeed in making them under laboratory conditions.

In Atlanta, primatologist Sue Savage-Rumbaugh taught Kanzi, a bonobo, how to make flint flakes so as to cut a rope tying up a food-containing box. At first, Kanzi could only make sharp pieces that were too small (less than an inch [2.54 cm]) to cut through the rope. One day, after four months of experimenting, Kanzi sat down gazing at the flint fragments. He then got up and, with all his might —and a bonobo is three times as strong as an average man—threw a flint stone at the cement floor. The stone shattered in many pieces from which he chose one of the right size.

In the spring, the experimenters moved outdoors, on soft ground. Kanzi again struck flintstones against each other, without much success. One morning however, he put a stone on the ground, stepped back, aimed and threw another stone at it, with smashing success!

As far as language is concerned, many species have shown unexpected skills. The American primatologist Penny Patterson taught sign language to Koko, a female gorilla that she treated as her own child. Koko commonly used up to 500 signs and recognized 500 others. Another chimpanzee, Washoe, is renowned for her language skills: she spontaneously combined the signs for "water" and for "bird" when she first saw a swan.

A spontaneous response to an unforeseen situation is the mark of a keen intelligence; it goes far beyond the learned response, the simple conditioning dear to behaviorists.

We should remember that communication is a means of

exchange of information between a sender and a receiver by means of a specific code. Language is a system of communication which, in addition, represents the world for the receiver. In George Page's words:

> Language is an immeasurably more powerful tool (and it is a tool) for dealing with the world than "communication" is.[1]

Human language is undoubtedly unique! Nevertheless, those who insist on the uniqueness of human nature should not be too worried by the linguistic skills of animals. The famous linguist Noam Chomsky points out that even without language animals can achieve amazing things, which are nevertheless not the result of real "thought." That being the case, the sasquatch is not "thinking." Witnesses report groans and shouts. In an earlier chapter, I described vocalizations. Not much to go by! A scientific analysis of recordings of these vocalizations brings out, at best, similarities with those of chimpanzees.

However, I would like to mention an experiment carried out by E. Menzel in the 1970s. It illustrates the mysterious fashion in which animals share thoughts and emotions.[2]

Menzel worked with six young chimpanzees. He took one of them—let's call him No. 1—to an enclosure where there was either some food hidden, or a stuffed snake (chimpanzees, like people, are afraid of snakes). He then took No. 1 back to his playmates.

There was no sign of any communication by No. 1 to his friends about the presence of either food or snake in the neighboring enclosure. However, when the door was open, all six primates rushed straight to the food, some of them even faster than No. 1. When there was a snake, the apes stepped into the enclosure with bristling hair and holding sticks.

Will we some day finally learn how these apes "speak" to each other? We certainly keep discovering unsuspected abilities in animals, whether they be free or in captivity.

For example, a few years ago in Jersey (British Channel

---

1. George Page, *The Singing Gorilla*, p. 112.
2. As told by George Page in *The Singing Gorilla*.

Islands), one of the most famous residents of Gerald Durrell's zoo, Jambo, a splendid silver-back gorilla, made front-page news. His massive build, the elegance of his gait, the sparkle in his eyes, and his aura of peaceful authority accounted for the feelings of visitors who fell under the charm of this impressive and benevolent giant. My wife and I admired him and paid him a visit each year until his death in 1991, at the age of 31 years.

The incident that spread his fame to the four corners of the world clearly demonstrated his intelligence and presence of mind. A five-year-old child, having escaped from his parents attention, managed to climb the low wall surrounding the gorillas' enclosure, and fell into it. He lay there on the ground, unconscious, with fractured arm and skull. Jambo sat near the child, guarding it, keeping the other gorillas away.

After a few moments, the situation having stabilized thanks to the calming influence of Jambo, a paramedic rappelled down into the enclosure and rescued the child while a couple of zookeepers kept the other members of Jambo's family, the females and their offspring, at a distance. The incident enhanced the public's esteem for the lovable giant and contributed positively to the reputation of his species.

In the annals of peaceful relations between humans and animals, there are numerous examples of people benefiting from the help of their so-called inferior brethren. Jambo's exceptional behavior presents a striking example. A life-size bronze statue commemorates the event.

I will certainly not venture into a polemic on the ethics of keeping animals in captivity, or of laboratory experimentation, preferring George Page's viewpoint:

> when I say to people that I'm studying animal intelligence, what I really mean is that I'm studying the way in which animals learn about the environments in which they live and I'm studying the way in which animals solve the problems that they confront in their environments.[3]

---

3. George Page, loc.cit. p. 103.

Life-size bronze statue of Jambo at Gerald Durrell's zoo, Trinity, Jersey, Channel Islands.

As for the sasquatch, it has never been observed in laboratory surroundings. No investigator has been able to submit it to tests that could measure its level of creative ability. But during encounters in the Pacific Northwest, some people have found themselves for a few moments face-to-face with a sasquatch. It is easy to understand their fright. Some however add that they had the impression of being probed to the depth of their soul, a feeling reminiscent of that expressed by Frans de Waal in his book, *Bonobos, the Forgotten Ape:*

> when the lively, penetrating eyes lock with ours and challenge us to reveal who we are, we know right away that we are not looking at a "mere" animal, but at a creature of considerable intellect with a secure sense of its place in the world.[4]

---

4. Frans de Waal quoted in George Page, loc.cit. p. 16.

Chapter 25

Back to *Big Footprints*

It took me a while to figure out why I was uneasy about Grover Krantz's book. At first, I couldn't take enough distance to form a critical assessment. I was probably still influenced by his presentations at conferences, passionate and assertive. His power of persuasion was enough to mask for me one of the essential traits of his book. As sources for his facts and arguments, Krantz calls upon writers on the sasquatch, like Peter Byrne, John Green, Daniel Perez or Ivan Sanderson. A few scientists (Bernard Heuvelmans, John Napier, Vladimir Markotic) are also cited in his bibliography, together with some of his own articles. Overall, however, the area of expertise contributing to the book is practically limited to the circle of classical sasquatch investigators of the Pacific Northwest.

Krantz does not draw on the works of other renowned investigators, as if they had nothing to offer to throw light, however dimly, on the mystery of the sasquatch. He thus deprives himself—and the reader—of useful points of comparison and potentially fruitful hypotheses.

Here's how I would rather proceed, as exemplified by a short citation from a collection of discussions between biologists, archaeologists and anthropologists (question addressed to Yves Coppens, professor at the College de France):

> Can the study of the behaviour of the great apes help us understand the daily life and social interactions of *Australopithecines?*
>
> The Australopithecines were not yet human; the ecological niche that they occupied was no more important than, nor very different from that occupied today, for example, by the pygmy chimpanzees of western Africa or the Bonobos. We are thus justified in comparing the behaviour

of the latter to the behavior and the social life of the *Australopithecines*, as long as we do it with prudence.[1]

To me, this type of working hypothesis seems most promising. There are after all numerous publications about the great apes by experts like George Schaller, Diane Fossey, Jane Goodall or the Leakeys, to mention but a few, all written so as to be accessible to the general public. Krantz also writes:

> Gorillas, chimpanzees and orangutans can be ranked as about equal to each other, thus almost certainly representing the same intelligence level of their common ancestor. Since the Sasquatch is also derived from that same common ancestor, we can safely presume that there was no loss of intellect in any of its descendants; such a loss would be selectively disadvantageous. This alone would argue for a mental capacity at least on par with the living apes.[2]

This is precisely where a comparison, even a superficial one, of the characteristics of these primates and of their respective behaviors would have helped define the sasquatch's own capacities and the bounds of its intelligence. Having come to recognize this absence as the source of my uneasiness about Grover Krantz's book, I strove to find references likely to satisfy my curiosity, and I invite the reader to follow suit. Not only can it open up fascinating new discoveries and a deepening of knowledge, it will also free the reader from the limitations imposed by an excessive, perhaps even obsessive, attention to bigfoot.

Let us now close this critical parenthesis, which in no way diminishes the real qualities of Krantz's book. He states that the sasquatch is neither fully human, nor even half-human. However, it may still be classified among hominids, implying that it is a biped,

---

1. Yves Coppens, in Combes et al., 1994. *L'Homme, Origine et Destinée*, p. 13
2. Krantz, *Big Footprints*, p. 170. According to Krantz, the common ancestor of men and apes would have lived ten million years ago; Coppens puts it at eight million years.

a mode of locomotion which the study of fossil Australopithecines suggests originated about four million years ago.³

The Australopithecines range in stature, depending on gender and age, from that of modern man to that, for example, of Lucy, the small female *Australopithecus:* 43–47 inches (110–120 cm) and 56 pounds (25 kg). Their brain was equivalent to that of an ape and their body was undoubtedly hairy. Though bipedal, they were also good at brachiating, using their arms to cling to branches. Their remains found in East Africa put them at 10,000 miles (15,000 km) from North America; if they also lived in Southeast Asia, they would still have been 7,000 miles (11,000 km) away.

Krantz points out a number of problems: Assuming that *Australopithecus* was the ancestor of sasquatch, how did it cross the Bering Strait? How did its height increase? How did its hands change? According to Krantz, an increase in size would automatically eliminate active brachiation, but would not prevent the retention of long and powerful arms.

The species which follows Lucy, named *Australopithecus robustus,* was more massive: 59 inches (150 cm) tall, weighing from 90 to 110 pounds (40 to 50 kg). Krantz sees *robustus* as a plausible

On the left, the skull of *Homo erectus;* on the right, that of *Homo sapiens;* in the center, the skull of a Neanderthal, too close to modern man, according to Krantz, to be the ancestor of the sasquatch.⁴

---

3. Yves Coppens was a leader of the team that discovered, in 1974, three million-year-old Lucy, "the most beautiful of the prehumans of the Afar savanas of Ethiopia."
4. Krantz, *Big Footprints,* p. 184.

precursor of sasquatch. The last known specimens occur about a million years ago. Assuming that they survived to this day, they would have had as much time to evolve as *Homo sapiens*.

Krantz then considers *Homo erectus* as a potential sasquatch ancestor. This species occupied the Old World from 1.6 million years ago to the appearance (less than 100,000 years ago) of Neanderthal man. His brain was half-way in size between that of man and that of the apes. He made some simple tools, sometimes lived in caves, did not use fire and probably did not wear clothes. That's about all we know!

In any case, even if the cultural level of *Homo erectus* is below that of Neanderthal man, it seems difficult to admit that he might be the ancestor of the sasquatch—he would have had to lose a number of his more human-like aspects to evolve into today's sasquatch.

Neanderthals are the most recent known possible ancestors of the sasquatch; they lived less than 100,000 years ago. In stature as well as in presumed intelligence, they were fully human, overall apparently too developed to be a likely ancestor. Their fossils have been dated from 100,000 to 30,000 years ago. Krantz describes Neanderthal man as a biped, with broad shoulders and long arms, possibly inherited from brachiating ancestors. His medium height was 66 to 67 inches (167 to 170 cm); he was hefty, typically weighing 200 pounds (90 kg). His feet were bigger than ours.

Neanderthals had brains as large as ours; they lived in caves, used fire, and made stone tools. They probably made clothes out of the skins of beasts. Some specialists claim that Neanderthals are a subspecies of modern man, *Homo sapiens neanderthalensis*. Krantz states they were a separate species.

Their bones were found in western Europe, many in southwest France, in the valley of the Vézère, and also in central Europe and in the Near East, as well as further east at the Teshik Tash site (Uzbekistan)—still more than 6,000 miles (10,000 km) from British Columbia!

We recall that during North Hemisphere glaciations, the sea level falls, the Strait of Bering is exposed, giving genus *Homo* access to the dry lands of Beringia and the rest of the Americas. Modern man, a contemporary of the Neanderthal, is thought to have spread into the western hemisphere from north to south.

As to how such a migration might have taken place, Yves Coppens suggests that after reaching a threshold level, one or two families would leave the main tribe, somewhat in the fashion of the Greenland Inuit, to settle some miles away on a new territory. As this group's population rose, the process would be repeated: Yves Coppens calculated that even at a mere 30 miles (50 km) per generation, it would take only 15,000 years to go from eastern Africa to the extremities of Europe and Asia.

As Krantz also points out, it is not distance that's the main problem, but rather adaptation to the Arctic climate. An even greater problem is the contradiction between the large cranial capacity of the Neanderthal and that, supposed, of the sasquatch.

Measured cranial capacities of hominids are as follows:

Australopithecines: (450–550 cm^3)
Homo erectus (Java): (770–1,000 cm^3)
Homo erectus (Peking): (900–1,200 cm^3)
Neanderthal: (1,300–1,425 cm^3)
Modern Man: (1,200–1,500 cm^3)

and where does sasquatch fit in this list? Between *Homo erectus* and Neanderthal man? If there is a link between a high level of intelligence, with corresponding brain size, the making of tools and clothes, and the use of fire, then Neanderthal man would have had to lose all these advances to become a sasquatch. Without forgetting the funeral rites suggesting the existence of metaphysical concerns—in Henry de Lumley's words: "Burial is the major invention of the Neanderthals."[5]

However, there was not enough time for Neanderthals to evolve, or rather devolve in this manner. Krantz points out their rather short time span: 100,000 to 35,000 years ago.

So, in spite of dissenting views among some scientists, particularly Russians, it is clear that Neanderthals cannot be the parents of

---

5. Henry de Lumley, *L'Homme premier*, p. 143. The author also describes an elaborate settlement of Neanderthals on the Moldova site, in Ukraine: a 23–35-foot-long (7–10 m) dwelling with walls made of mammoth bones, containing up to 15 hearths.

the sasquatch. Who then? Krantz adds: "That Neandertals might have been the ancestors of a different kind of unknown biped in Asia is another matter"[6]

Ray Crowe's views as to the relationship between Neanderthals and sasquatches are of interest. On the very day that I am writing these lines, the morning mail brings in the bulletin of the Western Bigfoot Society, *The Track Record*, No. 100.[7] The editorial, by Ray Crowe, speaks of stone throwing, an art in itself, which requires 88 different muscles in men as well as in apes. Apes throw projectiles to scare away their enemies. Men throw them to kill, to hunt, for self-defense—or for conquest. It would appear that the growth of the human brain might be linked to increasing skill in aiming and throwing accurately.

Crowe draws on the example of the pre-Neanderthal Acheulean man, who lived 1.6 million years ago, and whose tools were discovered in 1872 at Saint-Acheul, a village near Amiens, France. Archaeologist Pascal Picq described these tools: besides the double-edged stones, there were also hatchets, burins, bolas, and a variety of fragments.

The hatchet, commonly known as "the common Acheulean hand-axe," might be a throwing rather than chopping tool, rotating in flight so as to fall sharp edge downwards to kill or wound game: basically, a deadly frisbee.

On that basis, Ray Crowe suggests a possible link between Neanderthals and their predecessors, and the sasquatch who, according to various eyewitness accounts by Indians and others, is a skilled and powerful rock thrower. It is said to prefer round concretions and to have stone throwing abilities well beyond those of modern man.

Speculations abound. Not surprisingly, since the ancestry and relationships between pre-Neanderthal hominid species is poorly known and full of contradictions. As Pascal Picq points out: "The work of paleontologists shows that the history of mankind between 1.5 and 0.5 million years ago remains very poorly known."

---

6. Krantz, *Big Footprints*, p. 185.
7. See Appendix 2.

Equally mysterious is the disappearance of Neanderthal man, a contemporary of our ancestor, Cro-Magnon man,[8] during tens of thousands of years. Did the Cro-Magnon eliminate the Neanderthal? There is no archaeological evidence for this. The two races lived side-by-side for a long time and may well have interbred, but the fossils are too rare to be able to tell. Did the Neanderthal succumb to an epidemic—a plague carried by rats for example? This hypothesis remains unverifiable.

Neanderthals survived for a long time in some areas, but their disappearance remains unexplained. Specialists now believe that it was a relatively slow process, and that it took a long time for the superiority of Cro-Magnon man to affirm itself completely. It would be a mistake to think of the Neanderthals as failures of human evolution. For thousands of years, they led the way, before eventually being overtaken.[9]

---

8. Cro-Magnon man, first modern men in Europe, discovered in a cave (cro) belonging to a Mr. Magnon at Les Eyzies-de-Tayac, Dordogne, France.

9. A conclusion summarized from the work of Combes et al., 1994.

Chapter 26

The Right Choice

According to Krantz, bigfoot ancestor would be a giant primate, *Gigantopithecus blacki* (*Gb* for short), of which three jaw fragments and about a thousand teeth have been found in northern Vietnam and southern China. These fossils go back, for the most recent, to 300,000 years; for the oldest, to a million years.

We recall that the great apes, or hominoids, distinguished from other monkeys by their shorter and higher face, and larger cranial capacity, spread out of Africa about 16 million years ago. In Pascal Picq's words,

> The Asiatic lineage of hominoids, also called the Pongidae family, experienced a broad expansion between 16 and one million years ago. It includes among others the *Sivapithecus* group (Siva's apes, Siva being one of the three main gods of Hinduism) of which fossils have been found in India, Pakistan and China. Among this group, there are giant species, as *Gigantopithecus,* the biggest known apes. The latter became extinct only recently, between one million and 500,000 years ago. The famous legend of the Yeti is probably based on these impressive but placid bamboo eaters."[1]

Nowadays, there are only five species of hominoids left. Of the Asiatic lineage of the great apes, the Pongidae, only the eponym ("pongo" means orangutan) remains, while of the African lineage there remain gorillas, chimpanzees, gibbons and humans. We are indeed simians and share a common ancestor with chimpanzees. However, as Pascal Picq adds, "there is no fossil available of the last common ancestor of gorillas, chimpanzees and humans."

So, while the great apes, or hominoids, of the last 14 million years present characteristics which make them likely ancestors of

---

1. Pascal Picq, *Les Origines de l' Homme,* p. 62.

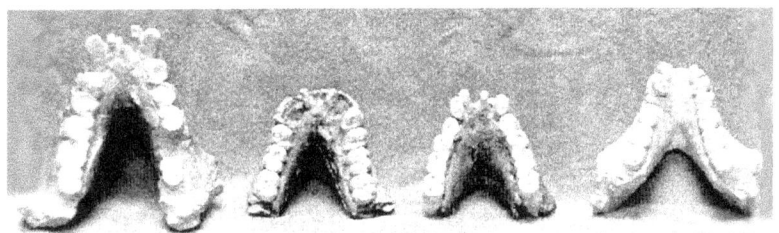

Gigantopithecus jaws. From left to right: adult male, young male, adult female—from China; extreme right, jaw of a female from India.[2]

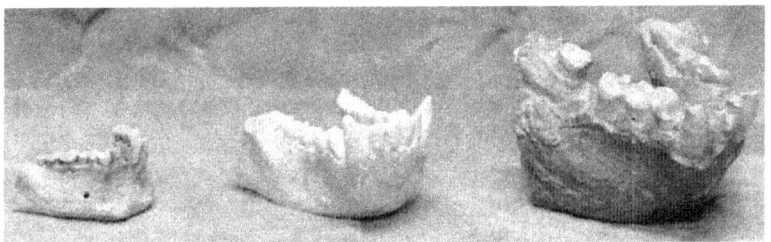

In the center, the jaw of a male orangutan; on the right, male *Gigantopithecus* (China); on the left, human jaw.[3]

hominids (the Australopithecines of 4 million years ago), in the absence of fossil evidence, only speculation is possible. For about 10 million years, the story of hominid evolution remains a big question mark.

There is clearly no reason why the Pongidae might not have continued to change, and in the case of *Gb*, might have endured for a long period. Based on Grover Krantz's reconstitution of its skull from the jaw and teeth of an adult male, *Gb* probably weighed over 900 pounds (400 kg). The following measurements, drawn from Krantz's work, are of great interest.

The height of the female was two-thirds that of the male, taller than any other known primate. No change in stature would be required to fit the characteristics of the sasquatch.

A detailed study of the jaw allowed Krantz to conclude that *Gb*

---

2. Krantz, *Big Footprints*, p.189.
3. Krantz, *Big Footprints*, p.189.

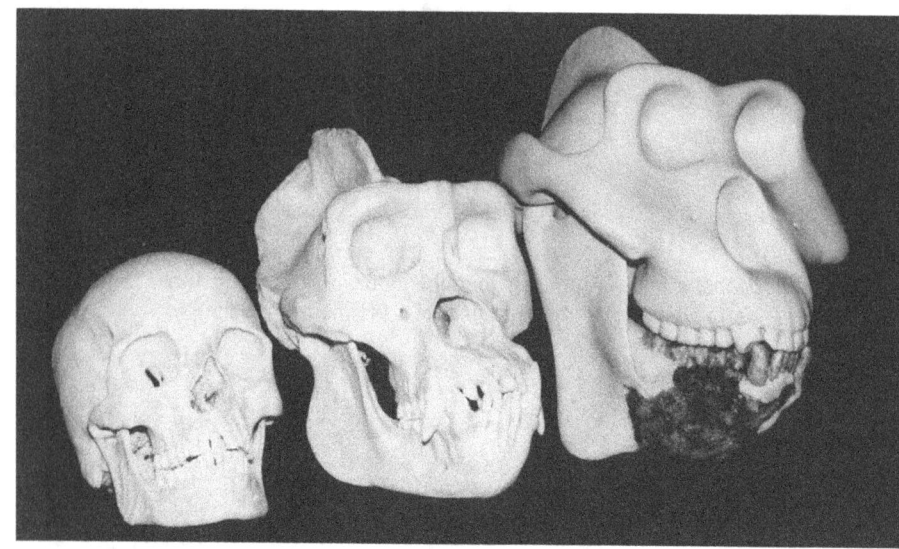

Dr. Krantz's *Gigantopithecus blacki* skull, right, compared with skulls of a gorilla and a modern human.

Artistic conception of *Gigantopithecus blacki*. *The creature* was over 6 feet, 7 inches (2.0 m) tall when it stood upright.

stood upright. The teeth are clearly hominoid and, without further details, suggest a brachiating structure. However, brachiation can be no more than a vestigial trait reflected in the width of the shoulders, since such a large and heavy animal could not readily swing from branch to branch.

Because of its longevity as a species, *Gb* was a contemporary of *Homo erectus,* probably without coming into contact with it, each occupying its own niche. *Homo erectus* used stone tools, hunted and trapped elephants, bisons and rhinoceros, but did not master fire.

An examination of the teeth of *Gb* reveals some details of its lifestyle:

A *Gigantopithecus blacki* model created by William Munns is highly sasquatch-like based on eyewitness descriptions.

> The dentition seen as whole is a grinding mill of the kind found in the Gelada Baboon and in Australopithecines; it is adapted to a diet of chewy plants: herbs and grains, requiring some manual preparation (cleaning, peeling…) and minimal use of the front teeth. The *Gigantopithecus* was probably entirely ground-based, as opposed to arboreal. How did it protect itself from predators? Without weapons or powerful canines, it is hard to imagine how its bulk, by itself, would be protection against attackers.[4]

In spite of its resemblance to *Australopithecus, Gb* is quite different from the genus *Homo,* particularly *Homo erectus,* and is unlikely to be one of our ancestors. It is more likely to have been the ancestor of the orangutan or of the sasquatch, this other extraordinary nonhuman primate.

Krantz thinks that sasquatch might have split from the Asian lineage three or four million years ago. If *Gb* is not the ancestor of sasquatch, one would have to look for another bipedal version of the

---

4. From *Encyclopedia Universalis,* 1955. We note that the sasquatch's main defense is its size; in North America, it is a perfect deterrent.

orangutan which would have developed somewhere in Asia before reaching America. However, adds Krantz, the evolution of an orang-like ancestor into as large a creature as the sasquatch would have taken a very long time. Nevertheless, such a transformation remains in the realm of the possible. However, Krantz would find the transformation of a Neanderthal or even of a *Homo erectus* into a sasquatch rather surprising, since it would require a greater degree of evolution in a shorter period of time.

When the remains of a sasquatch are found, what will it be called? Unless substantial differences are found, necessitating a new name, *Gigantopithecus blacki* will probably stand. Krantz suggests *Gigantopithecus canadensis*. In the meantime, one has to be satisfied with "bigfoot" or "sasquatch."

For the scientific community, sasquatch do not exist. Eyewitness reports and footprints are not sufficient evidence. However, numerous named and accepted fossils of shells are simply natural molds of animals of which there remains no substance. One might think that footprints would be just as valid. That is not the case, in spite of the growing number of samples gathered; Krantz himself owned quite a few.

At the end of his book, Krantz summarizes in his usual lapidary fashion the main points of his case:

1. Sasquatch shows every indication of being a perfectly normal species of wild primate, and is not connected with any paranormal phenomena. It is a legitimate subject of scientific investigation.

2. There is no reason to think that Sasquatch has superhuman, human, or even semihuman intelligence. Accordingly, we should treat them with the same consideration that we give to the great apes.

3. We have no indication that Sasquatch is an endangered species. Its population probably numbers in the thousands, and maybe tens of thousands. This is far from certain, however, and this idea might have to be revised when we know more.

4. It is not possible to prove that the species exists by continuing to collect sighting reports, footprint casts, and other minor evidence. Proper scientific study and possible protection will occur only when a type specimen is obtained.[5]

However, Krantz's bitterness, a feeling which he shares with many sasquatch hunters, often surfaces, especially when he speaks of the stubborn denial by the scientific community. In his view, the current consensus accepts the existence of the three-million-year-old fossils of *Australopithecus* while rejecting their giant modern equivalent: there is no room for thinking beyond the current paradigm. The Dutch scientist Eugene Dubois was the butt of similar scorn when he discovered *Pithecanthropus* fossils in Java in 1891. He had to bring his samples in and convince the authorities of the day to have look at them. Raymond Dart encountered a similar reception when he discovered the first *Australopithecus* in South Africa in 1924.[6] In both cases, notes Krantz, perseverance paid off.

---

5. Krantz, *Big Footprints,* p. 256.

6. Yves Coppens describes the problems encountered by Dart in his book *Le Genou de Lucy* (*Lucy's Knee*), pp. 125–135.

Chapter 27

Phylogeny

Phylogeny is the evolution of a genetically related group of organisms through time, as often represented by a family tree. Why be concerned with that, asked a friend: what is really fascinating for everyone is the mystery associated with the giant footprints left by the sasquatch and the enigma of its existence. Phylogeny only adds a needless complication to a phenomenon partly based on the imagination—why dig so deep?

I clearly felt that a scientific approach would threaten my friend's perception of the sasquatch. I pointed out to him that Krantz's work, his book and his other publications, compelled me to write about anthropology and anthropologists: one should attempt to position the sasquatch within evolution.

On the other hand, it also seems to me that the discoveries made by prehistorians, seen in museum exhibits, film documentaries, or illustrations in popular works as well as textbooks, are a powerful stimulus to the imagination. Besides, in the past 20 years, prehistory has experienced a renewed popularity expressed in a plethora of books, journal articles, and television programs, as well as by an increase in the number of archaeological expeditions and visits to sites of cave art.

Those who wish to delve deeper into the saga of evolution have to put up with the multiplicity of theories and the complexity of expert opinions. A specialized and specific vocabulary characterizes the distinctions introduced by paleontologists, as exemplified by this passage from Pascal Picq:

> The family hominidae includes two subfamilies: the paninae, to which belong today's chimpanzees and gorillas, and the homininae, of which the only representative is modern man.[1]

---

1 Picq, loc.cit. p. 68.

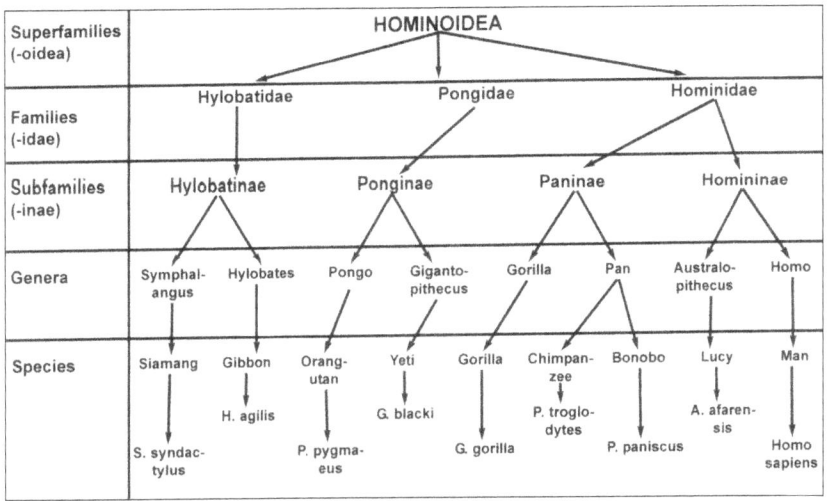

Hominoidea chart, after Pascal Picq in *Les Origines de l Homme*, p. 45.

Picq goes on to explain through a classification diagram which clarifies the relationships between extinct and current species. Families and subfamilies branch out from a common ancestor. Two centuries after Linnaeus, we find again that the chimpanzees are close relatives of man. The yeti also turns up as a survivor of *Gigantopithecus blacki*.

Family trees remain rather complicated for the uninitiated. Another scientist, Henry de Lumley, notes that "there are as many family trees as there are paleoanthropologists," and Yves Coppens, whose Human Phylogeny diagram is shown on the next page, adds, "the nomenclature of early Man is still rather confusing."

Readers of treatises on prehistory as well as visitors to museums are faced with an avalanche of special terms, often becoming obsolete as fashions evolve. This is why I have felt it necessary to sum up the broad lines of human evolution before returning to the enigma of the sasquatch.

First of all, let's note that humans (the genus *Homo*) appeared rather recently, about 250,000 years ago, as *Homo abilis*. Before the appearance of this first representative of the *Homo* lineage, there were a number of bipedal hominids, the Australopithecines, the most famous example of which is Lucy, a small female (4 ft [1.20 m]) discovered in 1974 in East Africa.

The Australopithecines, are rather close to the apes (the Greek *pithekos* = monkey or ape) but show some primitive human fea-

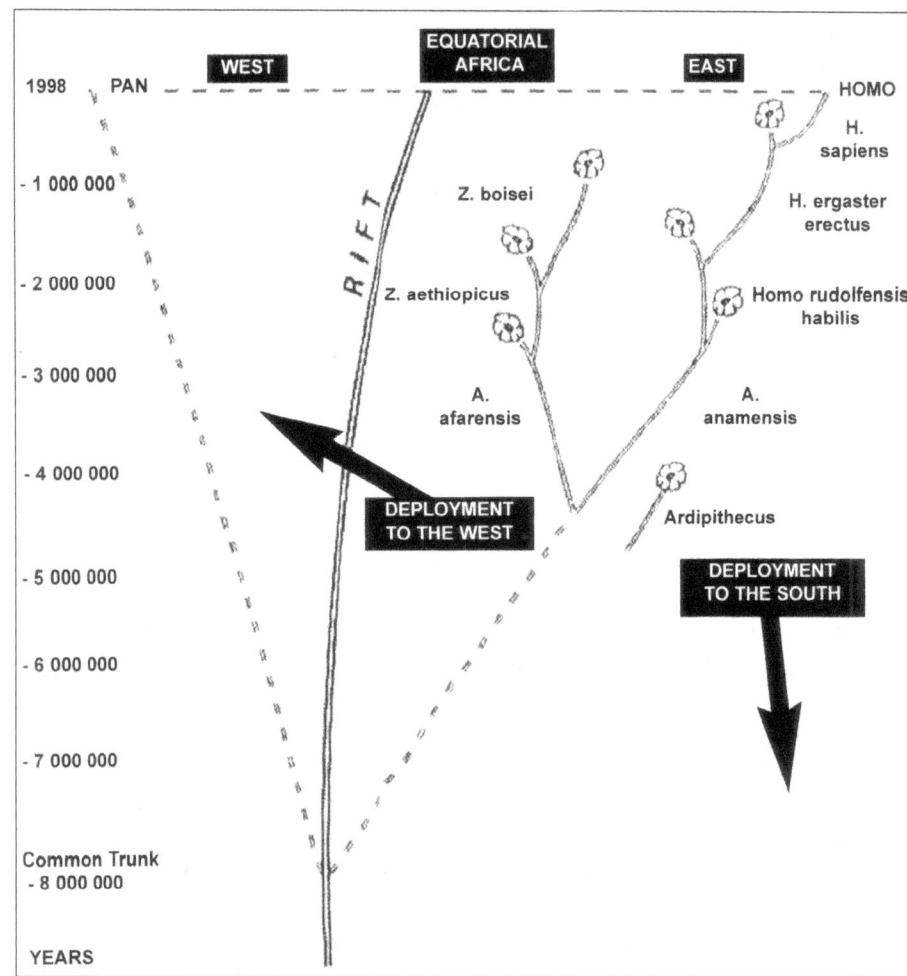

Human phylogeny chart showing the spreading of prehumans east and west of the African rift valley. "A" stands for *Australopithecus,* "Z" for *Zijanthropus* (a variant of *Australopithecus*) and "H" for *Homo.* (After Yves Coppens, in *Le Genou de Lucy,* p. 43.)

tures: they are bipedal and make stone tools. They are the first known hominids. They disappear from the fossil record about 3 million years ago, to be replaced by *Homo abilis,* followed in turn by *Homo erectus* its contemporary until about 1,800,000 years ago.

Finally comes a third species of *Homo: Homo sapiens,* modern man, about 100,00 years ago. Yves Coppens explains that modern man in contrast with prehumans (Australopithecines) spread far beyond the limits of their cradle, through Africa, and then Eurasia, probably through the Near East, but perhaps also across the Strait of Gibraltar, and much later Oceania, Australia and America.

We recall that *Gigantopithecus* survived as a species for a long time; the oldest jaws and teeth of this primate go back to at least a million years. They were still around at the time of *Homo erectus*, perhaps even sapiens, but probably did not share the same niche. Had they been in competition, said Grover Krantz, one species would have gained an advantage and outcompeted the other.

Gigantopithecus moved away from Asia towards North America and found a refuge there. With its massive hair-covered body, it is the only hominoid preadapted to the area and its cold winters, continues Krantz. With its imposing stature, *Gigantopithecus* is for Krantz by far the most likely candidate as an ancestor of the sasquatch.

One should not forget, however, that not very much is known about *Gigantopithecus*. Paleontologist Louis de Bonis has written about the unorthodox discovery of the first tooth in 1935 by a German scientist in a Hong Kong pharmacy, where "dragon teeth" were sold as "remedies." One of these was unusually large; after some searching, he found three others. Chinese scientists enquired as to the origin of these fossils, in caves of Guangxi (in the south of China). Unfortunately, the sediments within which the fossils were embedded were rich in phosphate and had long been used as fertilizers by local farmers, and thus widely dispersed. Nevertheless excavations brought up three mandibles and hundreds of additional teeth.

The 20-some mammals found with the *Gigantopithecus* remains were mostly carnivores. Louis de Bonis wonders: "Perhaps the giant primate was more of a prey than a predator?" In his book, *La Famille de l'Homme,* de Bonis reflects on the implications of the captivating drawing by Czech artist Zderek Burian, which shows a group of *Sinanthropus* (*Homo erectus pekinensis*), armed with chipped stones and spears, attacking a *Gigantopithecus.* Perhaps it was eliminated by the first humans 500,000 years ago!

De Bonis also states that there is some evidence that *Gigantopithecus* may be closer to man than to the chimpanzee.

It has been suggested that the gigantic Chinese primate might have survived as the yeti, a humanoid giant first reported in print in the 1420s and in recent times befriended in Tibet by an adventurous young Belgian journalist.[2]

In any case, the survival of a giant primate relative of man is evident for some and a matter of anticipation for many!

---

[2]. An allusion to the Tibetan adventures of well-known cartoon character Tintin, wherein he rescues a friend from an airplane crash with the help of the yeti.

Chapter 28
Interlude

As I write these lines, surrounded by recent works on human origins, buried in a pile of magazines—*Pour la Science, National Geographic, Science et Avenir, Science et Vie*—with their special issues, and drowning in a flurry of newspaper clippings, I am struck by the surging interest in the study of human evolution. Take for instance this brief article published in *L'Express* (December 14, 2000) summarized as follows:

> The discovery in 1974 of Lucy, the Australopithecine, took us back three million years; that of the bones of "millennium man" by French paleontologists Brigitte Senuet and Martin Pickford on the shores on Lake Baringo, in Kenya, goes back six million years. There is no doubt as to the ages: dating of the lava layers which surround the fossils has by now been mastered.

"Even though these hominids have many features in common with chimpanzees, there are signs indicating that they are already in the human lineage," says Martin Pickford.[1]

The lively debate about the transition from apes to humans continues.

The eddies of the stream of consciousness bring to mind two advertisements seen in the summer of 2000. The first praised the merits of an ecological—and expensive—voyage to Africa to bring participants in the proximity of gorilla families. The other described the program of a European competition of "prehistoric" sports, a

---

1. Martin Pickford, as quoted by M. Batiste, *L'Express*, December 14, 2000.

perfect occasion to witness the workings of the weapons of our Paleolithic ancestors, especially bows and spearthrowers.

New techniques open the door to the secrets of the past, brought to us by the fossils buried in the ground. Thanks to DNA analysis, the roots of family trees stretch far back into the past. A limitation is the small number of human fossils—only about 100 in the whole of Europe.

Bryan Sykes, a professor at Oxford University, has extended the work of the geneticists of the 1980s, who had already come to the conclusion that an African woman was the ancestor of all *Homo sapiens* about 150,000 years ago. Sykes looked for those daughters of the "African Eve" which were the ancestors of the European population. He found seven, the most ancient (that he called Ursula) being the matriarch of Cro-Magnon man, 45,000 years ago.

Beyond its promise as a tool for exploring human relationships and ancient migrations, DNA genealogy has raised various concerns. Social philosopher Jacques Ellul noted, "technical progress is not necessarily human progress, especially if it is not subject to democratic controls."[2] One can thus imagine insurance companies charging exorbitant rates to members of "families" less resistant to diseases; or enterprises eager to recruit "well born" and aggressive personnel.

Aware of the growing public interest in genealogy, Bryan Sykes created, jointly with Oxford University, the firm Oxford Ancestors: "The first organization in the world to offer a genealogical service based on genetics." Oxford Ancestors already has thousands of customers. Some of them have been quite surprised by what the tests revealed. For example, a Connecticut Italian American had traced his maternal ancestry back to 1820 to a Mexican woman living in Arizona. He was expecting Spanish ancestors. For lack of archival documentation, he turned to Oxford Ancestors:

---

2. Jacques Ellul (1912–1994) was a French philosopher, sociologist, theologian, and Christian anarchist. He wrote several books about the "technological society," and about Christianity and politics. The quote, originally from Ellul's "La Technologie ou l' Enjeu du Siècle," is from an article by Noël Mamère in "La Nouvelle Généalogie ADN," *Science et Avenir,* April 2001.

The analysis of my mitochondrial DNA (transmitted only by the mother) showed it to be most similar to that found in Nigeria or Ivory Coast. Once over my surprise, I carried on further research. I discovered that Mexico, like the United States, had imported many slaves from that part of Africa. I suppose one of my mother's ancestors was an African woman who lived about 500 years ago.[3]

As for paleontologist Yves Coppens, he found that he was not descended from one of the seven European "Eves," but from one of the 26 descendants of the African Eve who had spread over the world. Coppens states:

I was not looking for the seven female ancestors of Europe, they seem to me more symbolic than real. In my view, these Eves stand for the seven most favored streams of European settlement. The question which could bring together Brian Sykes expertise with mine is: what are the cultural trends which accompanied the migratory movements through pre-historic Europe?[4]

Social scientists clearly feel threatened by the scorn with which molecular biologists view their work. Archaeologist and prehistorian Marcel Otte protests:

Genetic genealogy is not an exact science...I find Bryan Sykes rather rash in selling—and it is indeed a business—these seven European Eves with such one-hundred-percent certainty.[5]

How far back can one reach? If any DNA found in the teeth of the *Gigantopithecus* were to be decipherable, could we find there traces of some of our ancestors?
What a comfort it would be to find, by digging further and fur-

---

3. *Science et Avenir,* loc.cit.
4. *Science et Avenir,* loc.cit.
5. *Science et Avenir,* loc.cit.

ther back, beyond millennia and geological layers, a solid foundation for our ancestry! But also, how naive to seek the reassurance of a direct lineage, free from the vagaries of history. It would be as if the latter could be fully replaced by the linearity of genetic genealogy, ignoring the ups and downs, ebbs and flows of biological and historical variability. Will we discover some day the amoeba from which I have evolved? The cosmic dust that gave rise to it—and beyond?

Chapter 29

In Other Lands: *Gigantopithecus* or Neanderthal?

In 1984, my friend and colleague Lauric Guillaud, the specialist of "Lost Worlds" literature, invited me to join him at a cryptozoological congress organized by Eric Buffetaut, professor of anthropology at the Jussieu campus of the University of Paris.[1]

Grover Krantz, Bernard Heuvelmans, Jacqueline Roumeguère-Eberhardt and Marie-Jeanne Koffmann were among the speakers. Their presentations excited my interest in "wild men."

Marie-Jeanne Koffmann (left) and Jacqueline Roumeguére-Eberhardt, 1984.

Marie-Jeanne Koffmann was born in France, the daughter of a Russian engineer who returned to the USSR when she was 16. She completed her medical studies in Russia and became a Red Army surgeon during World War II. She received many decorations and by the end of the war had reached the rank of captain. While on the Caucasus front she acquired a taste for mountain climbing and familiarized herself with the local people and their languages. Curious, but at first rather skeptical, about the stories of hairy humanoids in the region, she gradually became convinced of the reality of their presence. She spent many years in the mountains, under spartan conditions, searching for a proof of their existence.

Her keen eye, deep knowledge and openness of mind made a deep impression on me, a feeling that was only reinforced in our few subsequent encounters.

Mrs. Roumeguère-Eberhardt, an ethno-sociologist researcher at the CNRS,[2] had been splitting her time between France and Kenya. Her work had led her, quite coincidentally, to consider eyewitness

---

1. An account of the Paris meeting of June 9, 1984 is to be found in the ISC Newsletter, Vol. 3, No. 2, pp 5–6.
2. The Centre National de Recherche Scientifique (CNRS) is the principal French state research agency.

reports by Kenyans suggesting the presence of various types of humanoids, which she classified by types as X1 to X5.

Following the presentations of Koffmann and Roumeguère-Eberhardt, I first entered into my notebook this reaction:

> I was greatly surprised by the similarities between the investigations carried out by people who were so different. For years, Marie-Jeanne Koffmann has been investigating a humanoid (called Kaptar or Almasty) for which she has the deepest respect. The numbers of such beings have been decreasing significantly over the past years. The borders [of the USSR] adjoining Iran and Afghanistan are strongly guarded, but the Almasty (men of the forest, or hairy men) pay no attention and are often caught in barbed wires. Soldiers on duty had orders not to shoot on such inoffensive creatures. However, some infiltrators have attempted to cross the border disguised as "men of the woods," so that border guards now shoot on all Almasty. This is the reason why the number of hairy men in this area of the Caucasus has been reduced today to a mere dozen survivors.

As to Jacqueline Roumeguère-Eberhardt, she describes rather strange creatures. Type X1, for example, which looks like an ape, is hairy and rather primitive looking. Both it and its cousin X2 seem driven by curiosity when they intrude into dwellings, watching and sometimes even touching people during their sleep.

Both investigators expressed a deep respect for their respective "protégés." They feared that an influx of tourists would scare away these naturally wary creatures.

Marie-Jeanne Koffmann explained that the publicity having surrounded the latest expedition in the Pamir region[3] would proba-

---

3. The Pamir Mountain Range is located in central Asia, mostly in Tajikistan and Afghanistan. The highest summit is Ismael Samani Peak (24,590 ft [7495 m]). In 1958, the Presidium of the Academy of Science of the USSR created a commission for the study of the snowman. The Botanical Institute of the Academy, which had its headquarters in that area, was charged with organizing an expedition in the Pamir region; Marie-Jeanne Koffmann was its official physician. Some of the members of the Institute claimed to be aware of the presence of man-like creatures in the area. The hastily organized expedition was a failure. It was followed by less well equipped, privately sponsored expeditions.

bly bring a horde of adventurers on the scene. Under these conditions, the almasty will certainly flee the area.

As part of their presentation, both Marie-Jeanne Koffmann and her colleague showed slides. Koffmann's illustrated the Caucasus, which she had frequently visited, especially Daghestan. There were standing stones, a second-century chapel, churches, women weaving, sun-drenched fields, mountains and frozen waterfalls. Both women clearly were in love with their second homes, in the Caucasus and in Kenya respectively.

Answering a question, Marie-Jeanne said with some sadness that she would probably die without having found the ultimate proof of her quest—a live almasty.

My conclusion was that both women had found, through the intermediary of the almasty or of the men-of-the-woods of Kenya, a means of establishing a close relationship with the elements: air, water, earth and fire (one of the Xs would sneak into a house at night to steal fire). Living in a world inhabited by creatures—neither men nor beasts—which they had never seen, they were nevertheless obsessed by their presence, perceived as an enhancement of the beauty of these familiar regions.

I had also noted some disturbing elements. Jacqueline Roumeguère-Eberhardt quoted the words of some Kenyans who said:

> We live parallel lives and do not try to enter into contact with them (the Xs): they are much more powerful than us. We coexist, we meet them, but we do not try to follow them or communicate with them.

To which Roumeguère-Eberhardt added: "As an ethnologist, I believed that all that was part of Masai mythology. However, both aspects coexist: reality and myth."

Marie-Jeanne Koffmann's words echoed the same idea: "I am certainly not the one who projected a myth onto reality; on the contrary."

Grover Krantz expressed a similar view: "The myths do not bring strong evidence of the existence of the Sasquatch, but they do support it."

As to Dr. Heuvelmans, he thought that in America there had

been a weakening of the sense of observation, even among country folk. That is why bigfoot is described rather sketchily, while in Asia people provide a plethora of details (e.g., the shape of the chin, of the eyes) Heuvelmans reminded us that cryptozoology is concerned with unexpected animals, as reported by natives.

Receiving reports from shepherds, tea-leaf pickers and farmers of the Caucasus, Marie-Jeanne Koffmann was impressed by their consistency and by the accuracy of anatomical details described by "uncultured" individuals. Thanks to their remarkable skills of observation of nature—nature being in a way the diary of the shepherd—the country people described an anatomically logical sequence free from any fantasy: the resulting portrait was that of a "paleanthrope," a creature no longer a beast but not yet a man. For the shepherd, the almasty was a highly inquisitive creature, moving silently and handling objects without any noise. Its language was a muttering, but people claimed they understood what it said.

Why had the handful of surviving almasty—a handful scattered over thousands of miles of steep mountainsides—suffered such a catastrophic decline in numbers over the past 50 years? This was probably because of shifts in the lifestyle and attitude of the local populations—abruptly passing from feudalism to an industrial economy. The species slipped towards extinction, as a species might in such a particular period.

The cryptozoological congress concluded at the end of the evening, and I regretted at the time that Eric Buffetaut had not arranged for an informal gathering over a glass of wine. We had to leave still laden with questions that we had not had time to ask. Nevertheless, we had rubbed shoulders with the men and women who had learned a lot in their search for these elusive creatures, and had communicated to us their thirst for knowledge.

At a later date, Marie-Jeanne Koffmann pointed out the importance of the Caucasus Mountains as a habitat for the almasty. In that region, near the cradle of civilization, the humanoid population had gradually been pushed back by *Homo sapiens* towards arid mountainous areas that were literally encircled by humans. The similarity between humanoids and people brought about in the latter an ambiguous feeling of fear mixed with pity for the creature. Peaceful relations were established. Food and clothing were even offered to the humanoids. A natural sympathy arose towards females and their babies. These humanoids were smart enough to take full advantage of the proximity of human beings. According to Koffmann: "The kind of link between people and humanoids seen in the Caucasus is, to my knowledge, not to be found anywhere else in the world today."[5]

Emphasizing that she does not believe in the almasty, but rather that she knows that it exists, given the wealth of information that she has gathered (including casts of footprints), Marie-Jeanne Koffmann points out that the almasty, which some think might be a large anthropoid ape, and others, like Professor Porchnev, see as a relic Neanderthal, presents features which are more human-like than simian. Without jumping to conclusions, she thinks that the species belongs to the human family, or to a parallel and neighboring lineage.

---

5. Marie-Jeanne Koffmann, from *In the Footsteps of the Russian Snowman*, Dmitri Bayanov, Moscow, 1996, page 23.

Chapter 30

Professor Boris Porchnev

By his profession, Russian historian Boris Porchnev (1905–1972) would have seemed to be quite unrelated to the preoccupations of anthropologists, zoologists or anatomists. However his work led him to an interest in prehistory. For example, in 1955, he completed a "Study on the food resources of fossil Neanderthals that lived in a cave discovered by Soviet prehistorians at Techik-Tack, in Central Asia."[1]

At the end of 1957, A.G. Pronine, a hydrologist, told the press that he had seen a snowman in the Pamir, in the Baliand-Kyik valley. His description of the region—a valley filled with "berry-bearing bushes and a multitude of marmot burrows"—stimulated in Prochnev's mind an ecological stream of thought: such food resources would provide sufficient subsistence for hairy bipeds.

From then on, Porchnev focused on verifying the following question:

> Since the days of the appearance on Earth of Homo sapiens, had the Neanderthals quickly disappeared or had they just slowly degenerated? And what if by some chance they had not completely disappeared? It is from that time on that I became really interested in the snowman![2]

Porchnev accumulated an extensive documentation, as evidenced in his work "La lutte pour les troglodytes." Among others, he drew from the works of Prof. Vitali Andreievitch Khakhlov, a

---

1. Porchnev, B. "The struggle for the troglodytes," the first part of a joint work with Bernard Heuvelmans: 'L'homme de Neanderthal est toujours vivant" (Neanderthal Man is still alive), p.38.
2. Porchnev, op.cit. p. 40.

Professor Boris Fedorovich Porshnev (1905–1972).

doctor of biological sciences, and author of studies on compared anatomy and ornithology.

In 1907, while he was still a zoology student, Khakhlov was travelling towards the glaciers of Mouztau, near the border between the USSR and Chinese Sinkiang (Chinese Turkestan). His Kazakh guide told him about the existence of the "wild man." After enquiries among local inhabitants, Khakhlov sent a report of his research to two of his professors at his university—M.A. Menzbeer and P.P. Souchkine. While the first was quite skeptical, the second strongly encouraged him to pursue his investigation, so much so that in 1911 the young man set aside his studies for two or three years to explore the area of Lake Zaysan, seeking information about the *Ksy-gyik* (wild man).

In 1914 Khakhlov solicited a grant from the Academy of Sciences in St. Petersburg; in spite of Souchkine's support, himself an Academician, Khakhlov's request was rejected. He turned to the Geographical Society, but the war had just broken out and the Russian government had more pressing concerns.

Khakhlov had taken very careful notes of the descriptions offered by his informants, and he also expressed them in sketches, such as the one below.

Sleeping Ksy-Gyik, in the pose of a child: "An antediluvian man," according to Khakhlov.

Among other observations, Khakhlov noted that:

> The Ksy-Gyik seeks, in a manner of speaking, the absence of human beings. During the summer, when shepherds lead their flocks to the high pastures, they descend into the valleys; in the winter, they do the opposite. It is seen alone or in pairs, with or without offspring. Most of the sightings take place at dusk or dawn, or even at night, rather than in the day.
>
> They are not known to have permanent shelters, but some temporary ones are found here and there. The Ksy-Gyik's food includes roots, shoots and berries, as well as birds' eggs, lizards and turtles. Most of the diet is made up of small rodents living in the mountains or in the sandy deserts."[3]

Porchnev went searching for Khakhlov's original manuscript, which he found in the archives of the Academy of Sciences. It had landed in the historical-philological section, under the label: "Notes without scientific importance." In 1914, the author had noted:

> The content of these stories, gathered directly from eye-witnesses, would suffice in my opinion, to regard them as more

---

3. Porchnev, op.cit. p 55.

than mythological, or simply imaginary. The reality of the existence of such a Primihomo asiaticus, as one might name it, can no longer be in doubt.[4]

In 1937, south of Lake Zaysan, Kazakh horsemen brought to a detachment of the Soviet Army a wild man they had captured. Marshall P.S. Rybalko gave a detailed description of the strange being: "It looked like a man or, one should rather say, a fossil ape-man."[5] Unfortunately, the creature did not survive its handling and died after eight days; its rotting carcass was quickly discarded.

It is noteworthy that the zoologist, the Kazakhs, and the officer all came up with a similar description: an ancient Neanderthal-like being, unfortunately not to be found in any textbook.

Porchnev reminded his reader however that in the 18th century, Linnaeus, in his *Systema Naturae*, had classified within the genus *Homo*, not only *Homo sapiens,* but also *Homo ferus,* the wild man, and *Homo troglodytus,* the cave man. He had also identified the fundamental differences between the latter and modern man:

The absence of articulated language (mutus = mute);
The abundance of body hair (hirsutus = hairy);
The ability to walk on all four (tetrapus = four-legged).

According to Porchnev: "Linnaeus did not think that these traits were enough to place such creatures outside the genus Homo."[6]

In summary, these creatures differed from modern man through their lack of language *(mutus)* and from the apes *(tetrapus)* by their bipedal stance. They were higher primates with vertical posture. Although clearly in decline, these neanderthalians had not completely disappeared. Porchnev mentioned the discovery of a few Neanderthal skeletons in Russian territory: near Irkutsk, in Karelia, in the Moscow district, and on the Dnieper River. More abundant finds were made in the North Caucasus.

Porchnev brought up as evidence, in addition to fossilized

---

4. Porchnev, op.cit. p 59.
5. Porchnev, op.cit. p 64.
6. Porchnev, op.cit. p 103.

bones, various likenesses carved in bone or stone, and more recently sculptures decorating churches and cathedrals. He also drew from ancient writings, such as the Kirghiz epic of Manas[7] (mid-ninth century):

> Have you heard of the countless marvels
> Along the road we shall follow?
> Who has heard of the real wild camel?
> Who has heard that there are real wild men?

Porchnev also quoted from Maupassant's novella *La Peur* (1884), where the author relates the frightening encounter made by his friend Turgeniev, in his youth, while hunting in the forest, in Russia. At the end of the afternoon, after a long hike, he had gone for a swim in a river. Suddenly, he felt a hand on his shoulder. Turning around, he saw a creature, half-woman, half-monkey.

> Two nameless features, no doubt breasts, floated in front of her, and an extravagant head of hair, matted and reddened by sunlight, surrounded her face and floated on her back.
> The young man, scared witless, swam ashore and leaped with all possible haste into the woods, without even thinking about his clothes and his gun.
> The frightening creature followed him, running as fast as he could, groaning as she went.
> In flight, nearly exhausted and stiff with terror, he was about to collapse when a child who was shepherding goats nearby came up with a whip. He struck the horrible woman-beast which ran away screaming with pain. Turgeniev saw her disappear into the foliage, looking like a female gorilla.[8]

---

7. A patriotic work recounting the exploits of Manas and his descendants and followers, who fought against the Uyghurs in the ninth century to preserve Kyrgyz independence. (Wikipedia).
8. La Peur was published in *Le Figaro,* July 25, 1884. Not to be confused with a tale of the same title published in *Le Gaulois,* October 23, 1882.

Few of the many reports gathered by Porchnev, whether ancient or recent, possess the literary quality of a Maupassant tale. However, their very lack of literary pretention and their monotonous simplicity are in some way a guarantee of authenticity.

In 1958, after gathering extensive documentary material, Porchnev set in motion new research activities. Following his studies in biology, and doctoral degrees in history and philosophy, he had achieved an important official position, which opened doors, but also exposed him to the scorn of his colleagues. However, the president of the Academy of Sciences of the USSR welcomed his suggestion: the creation of a commission to study the question of the snowman.

The commission had to fulfill two objectives before launching an expedition into the Pamir mountains. First of all, to gather all information available, from the whole world, on the subject; second, to function as scientific advisory body to a mission aiming at identifying the creature reported by hydrologist Pronine. The expedition was organized on a vast scale and with great thoroughness: rubber raft, speedboat, police dogs trained "using chimpanzee urine as a stimulant."

However, the membership of the scientific team was somewhat surprising. It included two ornithologists, a specialist on seals, another on bats, and an entomologist. The attitude of the expedition's leader, K.V. Staniukovitch, reflected the lack of interest of most of the participants in the snowman. The leader was obsessed with completing a geobotanical map of the Pamirs. All participants were there primarily to increase their knowledge in their own scientific specialty.

Of course, to Porchnev's regret, the expedition leader decided to spend the summer in the Pamirs, at the time when the wild flowers were in bloom. In the spring, the valleys are still covered with snow and tracks are noticeable, which is not the case in the summer. As Porchnev commented:

> The whole time was spent on ascents and descents on horseback among rock piles and screes, cruises in canoes or motorized rafts on a 60 sq. kilometer (24 sq. mile) lake, (hitherto never traveled upon by man) and night vigils in camp sites hanging off the rare edges of steep shores. While

the whole trip was a marvelous, unforgettable experience, it did not remotely relate with the goals of the expedition."[9]

We can imagine that Porchnev's disappointment was somewhat attenuated by the presence of Marie-Jeanne Koffmann, who had offered her medical services to the expedition. Porchnev was extremely complimentary of her work, praising above all her vitality. She often ventured alone in search of relic paleanthropes, traveling in a jeep, on a motorcycle, on horseback, and even on foot. Over the years, she collected an impressive corpus of information. In 1962, she moved her base of operations to Kabarda-Balkaria (in the northern Caucasus). As a lone woman in a Muslim region, she succeeded, mainly thanks to her knowledge of local dialects, to win the confidence of the natives, whose eyewitness reports she meticulously noted.

Whimsically, Porchnev wondered, after all, whether all this was only folklore, the timeless warp upon which fairytales are woven? As we shall see, many documented events do seem to belong to the realm of imagination.

---

9. Porchnev, op.cit. p 87.

Chapter 31
Zana the Ogress

Some stories certainly sound like tales of fairies and witches. Such, for example, is that of Zana, described by Porchnev and later by Bayanov.[1]

One day, in Abkhazia (eastern Caucasus), a group of hunters captured a dark-skinned female. She was struggling furiously and they tied her up. Sold a number of times, she eventually became the property of a local noble, Edghi Ghenaba.

She was locked up in a solid paddock, where she behaved like a wild beast. She dug a sleeping hole in the ground. For three years, she was fed by throwing food on the ground. As she became domesticated, a small sun-screened enclosure of intertwined branches was built for her near the house. After a while, she was allowed out without a leash.

She could not stand heated rooms and slept in her hole, under the sun screen. She bathed, even in

Artistic conception of Zana by Brenden Bannon. See the color section for the full image.

the winter, in a freezing stream that still bears her name. In the summer, to cool off, she lay down in a puddle next to the cattle.

She chased dogs away by throwing heavy sticks at them. She enjoyed breaking stones by hitting them against each other. She tore off the dresses she was given to wear, preferring to go naked. Eventually she got used to wearing a loincloth.

Sometimes, she would come into the house, but the women

---

1. Bayanov, D., *In the Footsteps of the Russian Snowman*.

were afraid of her, for she would sometimes bite when she was angry. Her master, who was enormously strong, could make her obey. She could run faster than a horse. With a single hand, she carried a 180-pound (80-kg) bag of flour from the water mill to the village. She climbed trees to pick fruits, and ate grapes by the bunch. She could drink large quantities of wine, after which she fell into a kind of alcoholic stupor.

Although she never attacked children, they were afraid of her, and parents in the area would threaten their brood with beckoning the " ogress."

An ogress she may well have been, considering her gluttony: she gobbled up, with her hands, everything she was offered, with a strong preference for meat and cornmeal.

Zana was taught to perform some simple tasks: turning a handmill, carrying firewood and flour bags, fetching water, pulling her master's boots. The best she could do was to light a fire with flint and tinder. She did not manage gardening, or riding a horse.

In the village of Tkhina, where she lived for some decades, Zana never learned a single word of Abkhazian. She uttered inarticulate sounds or incomprehensible howls. However, she was sharp of hearing and would come when her name was spoken.

Her face was frightening and large, with sharp cheekbones and rough features. Her nose was both upturned and flattened. The lower part of her face stuck out, like a muzzle, with a big mouth and large teeth. Her forehead was low, with thick eyebrows, and her eyes had a reddish tinge. Her expression always remained that of an animal, without a trace of humanity.

Zana lived for many years without any change in physical appearance: no loss of teeth, white hair or loss of muscle tone. Her skin remained black or dark gray, covered with reddish black hair.

Zana gave birth, without any help, to many children which she immediately washed in the streambed. These small hybrids all died and soon the villagers took her newborns away from her to raise them. Thus survived two sons and two daughters who, in spite of some physical and mental peculiarities, turned out to be able to fit in Abkhazian society. The younger son, Khwit died in 1954. Rumor had it that Edghi Genaba was the father of those children.

When Porchnev visited the region, in September 1964, together with the archaeologist V.S. Orelkin, he interviewed local people who had known well both Khwit and the younger girl, Gamasa. Porchnev described them:

> Both were sturdy individuals, with dark skin and some negroid features. They had inherited hardly any of the Neanderthalian features of Zana. The human characteristics had dominated and erased the other component of their heredity.[2]

Khwit

Porchnev returned to Abkhazia three times, in vain, to discover Zana's grave. In his last attempt, in October 1965, he exhumed Gamassa's bones: they presented significantly neanderthaloid characteristics.

After Porchnev's death, Igor Bourtsev led three expeditions to Abkhazia, in 1971, 1975 and 1978. Khwit's skull was dug up and studied in Moscow by a pair of anthropologists who detected both modern and primitive characteristics.

One might regret that investigators did not manage to find Zana's remains. On the other hand, she deserves to rest in peace.

One more item: in 1962, zoology professor Machkovtsev dug up an arrowhead, chipped from a pebble, on the hillside where Zana loved to wander. Strangely enough it was of Mousterian type. A coincidence? The Mousterian culture—from the Mouster cave, in the Dordogne district of France—is characterized by new stone chipping techniques from flint blocks and pebbles.[3]

The stirring story of Zana brings us back to the hypotheses about the disappearance of the Neanderthals. Against the holders of

---

2. Porchnev, B. loc. cit. p. 175.
3. Mousterian is a name given by archaeologists to a style of predominantly flint tools (or industry) associated primarily with *Homo neanderthalensis* and dating to the Middle Paleolithic, the middle part of the Old Stone Age. Mousterian tools that have been found in Europe were made by Neanderthals and date from between 300,000 BP and 30,000 BP. In Northern Africa and the Near East they were also produced by anatomically modern humans. (Wikipedia).

Igor Bourtsev examining Khwit's skull at the grave site. The inset shows a frontal view of the skull.

a sudden disappearance, the specialists of the St. Césaire Man[4] showed that the remains they found in the upper paleolithic levels of that site clearly demonstrate a cohabitation in time of Neanderthals and modern man. For millennia, the two populations exchanged technological innovations. Did they also have biological intercourse? Zana's story suggests the possibility of hybridization.

For some, hybridization is out of the question, so that in the limit, Neanderthal man would be a different species, completely separate from *Homo sapiens*. For others, the recent discovery in Portugal of parts of the skeleton of a child born 24,000 years ago would indicate the existence of a hybrid population. The skeleton

---

4. St. Césaire, a village in the Charente area of France where Neanderthal remains have been found and dated at about 36,000 years ago.

has both neanderthalian and modern human characteristics and belonged to an individual born after the disappearance of the last Neanderthals (but, had they really disappeared?). Cross-fertilizations would thus have given rise to a population presumed to have lived isolated for 4,000 years, i.e., about 200 generations, without noticeable influence on the future of mankind.

Homo sapiens achieved dominance by his expertise in exploiting his environment, by his breeding success, and by the brutal elimination of his rivals. However, it is probably the development of symbolic thinking that has bestowed to modern man the dominion of the Earth. Neanderthal was left behind. Although its brain closely resembled that of sapiens, its symbolic practices limited its ability to act on the world around it.

Does natural selection explain the surprising abilities of modern man? Here's how I would summarize what Ian Tattersall has to say about it: The best way to describe evolution is to say that it is opportunistic, that it merely exploits or rejects possibilities as they appear; these possibilities in turn may be favorable or not, depending on the environmental circumstances (in the widest possible sense) of the time. The process does not obey any secret directive, there is nothing inevitable to it, and it may reverse itself if the environment, ever variable, were to change.

If Darwinian selection has nothing to do with the creative process, one must conclude that *Homo sapiens,* under the influence of some stimulus, took a veritable quantum leap. What could have been the nature of the stimulus that provoked such extraordinary cognitive progress? It's anybody's guess.

We should not forget about the Neanderthal's great talents, undoubtedly due to its great powers of intuition. Its intuitive reasoning allowed it to prosper for a time. The current fascination with Neanderthal man stems in good part from our unconscious admiration for its emotional brain. That part of the brain, called in turn the ancient brain, or the olfactory or visceral brain, and later the limbic system, is the favored seat of sensations and emotions. It might also be the seat of mythical thinking, which may arise, as Bernard Heuvelmans suggests, at the level of the limbic system, appropriately nicknamed our emotional brain.

Reminiscing about Neanderthal man awakens feelings buried deep in our ancient brain. Its adaptation to the natural world contrasts with the complexities or the simple-mindedness of today's prevailing myth, that of science, in the sense of "All-Powerful Science," inebriated with the latest successes of information technology or genetic engineering, ready to set the gene at the focus of life, while ignoring the living organism. Not of course science as it proposes explanations to clarify specific phenomena, or to demonstrate the effects of some elaborate mechanism, for example that of nocturnal vision, which we discuss below.

Chapter 32

"Slanted, bloodshot eyes" ... frequent blinking

The reader must certainly have noticed that remarkable detail about Zana: "her eyes had a reddish tinge." Marie-Jeanne Koffmann looked into this trait, often mentioned in eyewitness reports from the Caucasus. Besides this reddish glimmer of the eyes, often mentioned were "strange periodic movements, unusual blinking of the eyelids." These traits also show up in reports on the sasquatch in the USA.

Dr. Koffmann presented a scientific study on this phenomenon.[1] To provide a concrete demonstration of her modus operandi and of the methods used by serious cryptozoologists, I have taken the liberty to summarize the results of her work.

The author starts by recalling the uniqueness of cryptozoological thinking, which:

> is characterized, inter alia, by careful attention to information, recent as well as ancient, arising from sources quite remote from science, often illiterate: shepherds, hunters, loggers, warriors, etc.,... in faraway countries.

The principal methodological tool consists in "interviewing local people." However, it is necessary to subject this unusual method to:

> an elementary critical analysis, as to the contents of the story, its repetitiveness, the diversity of sources, the plausibility of the individuals described, how they fit into their

---

1. Marie-Jeanne Koffmann, "Relic humanoids of the Caucasus. The use of comparative anatomy and physiology in the authentification of eye-witness reports in cryptozoology." (In French) First European Conference on Cryptozoology, Rome, 27–29 March 1999.

environment, the personalities of the witnesses, as well as their number. At the next stage, in order to support his hypotheses, the investigator must consider the geomorphological and paleontological history of the region; the history of its fauna as well as that of its colonisation by man; the works of ancient authors; the reports by travelers, merchants, naturalists, and missionaries, etc.,...of centuries past; the archives of various administrative bodies: religious, civil, and military; archaeological data; works of art, mythology, folklore, linguistics, toponymy, and many other related sources.

Marie-Jeanne Koffmann thus defines, in a few words, the basic principles of cryptozoological research, which turns out to be so attractive to the curious mind. She also applies these principles to the question of the almasty's reddish eyes.

She starts by reviewing the anatomy of the human eyes. The human eye is wrapped in three tunics of tissue:[2]

—an external fibrous tunic, the sclerotic coat, or sclera, the forward third of which is the transparent cornea;

—a vascular tunic, the choroid, which forms the iris in the front of the eye and which lets the light in through the pupil;

—a nervous tunic, which includes the retina, the inner layer of the eye made up of ten layers, one of which includes rods and cones which are the light receptors. The some 7 million cones are responsible for color vision and operate only in relatively bright light. The rods (about 130 million in the human eye) are extremely sensitive to light but do not provide a sharp image and do not detect colors.

Marie-Jeanne Koffmann continues:

---

2. An excellent overview of the anatomy and functioning of the human eye is to be found at <http://users.rcn.com/jkimball.ma.ultranet/BiologyPages/V/Vision.html>.

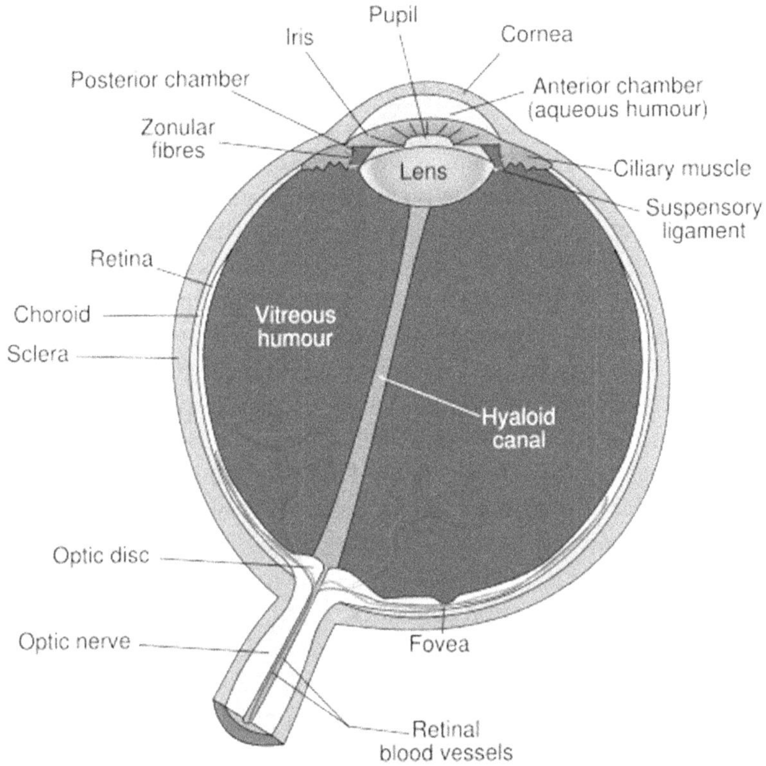

Pupil
Iris
Cornea
Posterior chamber
Anterior chamber (aqueous humour)
Zonular fibres
Lens
Ciliary muscle
Suspensory ligament
Retina
Choroid
Sclera
Vitreous humour
Hyaloid canal
Optic disc
Optic nerve
Fovea
Retinal blood vessels

Above: A contemporary diagram of the human eye. Right, the eye illustration from Marie-Jeanne Koffmann's material.

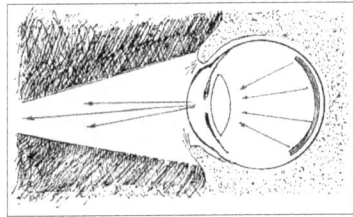

Strangely the vagaries of phylogenesis and embryogenesis have inverted the layers of the eyes of mammals, so that the receptive layer of the retina, holding cones and rods, is at the back of the eye, turned away from the incoming light, which must travel though the other nine layers before reaching the sensitive receptor cells. At that point, the light is held partly back within the retina by a layer of melanic pigments, the dark tapetum, but some of the rays leak through to be absorbed in the choroid or even in the sclera.

To improve their night vision ability, nature has provided nocturnal animals with an additional layer, the tapetum lucidum ("the bright carpet"), a bluish or white reflective layer, behind the retina, in the choroid layer. It is composed of high density collagen fibers (in ungulates) or by multiple layers of flat plates filled with guanine silver crystals (in carnivores), the same substance which is responsible for the reflectivity of fish scales. The tapetum lucidum functions like a mirror, reflecting light back to the receptor cells...

It so happens that in some cases the color of the light reflected from the tapetum varies according to sex, within the same species. In Russia's southern steppes, wildlife zoologists assess the sex ratio of herds of saiga antelopes by the color of their eyes: red for males, pale blue for females, pink for juveniles.

The natural history lesson continues:

Like all diurnal animals, the higher monkeys (with the exception of the Douroucouli or Nyctipithecus, a nocturnal monkey of South America) and man, do not possess a tapetum lucidum. However, there are occasional reports of superior night vision in some people whose eyes have a red reflection in the dark.

Koffmann then elaborates with a complex description of the role of rhodopsin, a blood-red protein pigment, in night vision.

What is important in all this is that people possess a rather poor night vision, much worse than that of monkeys. But there exist the means of getting around this deficiency through off-center vision—looking at an object slightly sideways—and by scanning slowly across the field of vision while blinking.

According to Marie-Jeanne Koffmann:

It is exactly because the humanoid (Almasty) does not possess the retina of a truly nocturnal animal, but a more typically human retina, that it has to adopt a scanning strategy to improve its night vision, the very same that people resort to. The transition to a nocturnal existence is thus probably a recent development in the history of the species.

Passing from a diurnal to a primarily nocturnal way of life, the almasty would have certainly benefited from having a vision adapted to the change. An improvement in its night vision would have been an advantage contributing to the survival of the species according to the criteria of natural selection. However almasty did not have the time required, since an enhancement in visual acuity would have required thousands and even millions of years.

In concluding this chapter, I am reminded of these words by Dmitri Bayanov:

> The study of relic humanoids is revealing not only of their reality but, in general, of a new kind of reality. Since humanoids differ from both the great apes and Homo sapiens, their study constitutes a legitimate branch of knowledge that deserves its own name. Thus was born hominology as a branch of primatology, filling the gap between zoology and anthropology, in full concordance with Darwin's thinking."[3]

---

3. Bayanov, D. *America's Bigfoot: Fact Not Fiction*, p. 9.

Chapter 33

A Few Years Earlier

Let us step back a few years to the days when Marie-Jeanne Koffmann was spending months on end camping in the Caucasus, in the 1970s.

The wild man was in the news on both sides of the Pacific. It was in December 1971 that independent investigator René Dahinden arrived in Moscow bringing, like Santa Claus, photos, footprint casts, recordings of interviews with Patterson, Ostman, and Gimlin, copies of John Green's works, documents copied by Dahinden himself as part of his own research and, best of all, a copy of the famous 1967 Patterson/Gimlin film.

Thanks to his Russian friends, Dmitri Bayanov and Igor Bourtsev, Dahinden made numerous presentations and film screenings. His directness, his sense of humour and his Swiss-accented English made him popular. The Anthropological Institute ignored him, but other groups welcomed him, especially the Institute of Scientific Research on Prostheses and the Construction of Artificial Limbs. Up to a hundred of their specialists and technical experts viewed the Patterson/Gimlin film. Then a half-dozen or so senior administrators also viewed it repeatedly. Dahinden understood that those people, in contrast to anthropologists and zoologists, did not risk their reputations. They were examining a mode of locomotion and movement; the exact nature of the creature was of little concern. The director of the institute considered that the film had enough clear frames showing the motion of a human-like being for a detailed scientific study to be deemed of real interest.

René Dahinden in Moscow.

Igor Bourtsev (left) and Dmitri Bayanov in 2002. (Photo, Daniel Perez)

The intense efforts of Dahinden and his Russian friends were to stimulate genuine interest at the same time as they triggered derision and even scorn from part of the Soviet public. One may recall the political climate that reigned at the time! Dahinden's visit was a breath of fresh air, an occasion for the creation of friendly bonds between the adventurous Canadian visitor and his Russian hosts. It was finally possible to exchange views with visitors without constraints at a time when it was forbidden to entertain any written correspondence in scientific, cultural or economic domains without the permission of some official authority. For example, Bayanov needed the official imprimatur of the director of the Darwin Museum, in Moscow, for written correspondence. Fearing a refusal, he preferred to distribute his work sub rosa, using underground communication channels.

In North America, on the other hand, reigned a widely envied openness. For example, Marjorie Halpin, the curator of the Vancouver Anthropology Museum, and her colleague Michael Ames, a professor of anthropology, convened a symposium entitled "Manlike Monsters on Trial" at the University of British Columbia, in Vancouver, May 8–13, 1978.

The organizers' intention was to foster an interdisciplinary debate on the biological and ethnological aspects of humanoids.

"One of the objectives of this meeting is to bring this important domain into the fold of academic research," said Ames. Hoping to attract scientists from everywhere, and aware of the work of the Russian investigators, they issued invitations to Dmitri Bayanov, Igor Bourtsev and Marie-Jeanne Koffmann. Unfortunately, the bureaucrats of the USSR Academy of Sciences did not allow their participation; their contributions had to be read on their behalf in absentia. The meeting received favorable press and the conveners gathered the contributions in a 336 page volume entitled, *Manlike Monsters on Trial*[1] It has become a classic for anglophone cryptozoologists.

In his presentation, entitled "What is the Sasquatch?," John Green pointed out the similarities between the sasquatch and the creatures described by Porchnev, noting that the main difference lay in their stature; the Russian-described hominid being neither taller nor heavier than normal man. Green stated that he believes the Russian hominids are neither "missing links," nor "near-humans." They have evolved along different, divergent paths relative to *Homo sapiens*, and lack his effective brain, his long use of tools, and his skills at verbal communication, which allow him to cooperate effectively with his brethren.

Furthermore, Green stated that the sasquatch and the almasty followed opposite paths, but not that of the great apes. In contrast to the latter, they can swim, they possess enhanced night vision and they can endure a variety of climates. Their adaptability makes them a species—perhaps not a single species—able to survive in many areas of the world. It is from that perspective that they resemble man, but only from that perspective, since their adaptation has been limited to their physical abilities.

John Green's conclusion was that if humanity is to be defined by possession of a vertical stance, then these creatures are human. If on the other hand, it is the nature of the brain that distinguishes *Homo sapiens* from the animals, then these creatures, almasty and sasquatch, are animals—apes that stand erect—and nothing more.

R. Lynn Kirlin and Lasse Hertel, specialists of signal processing, presented the results of their analyses of recorded vocalizations

---

1. Halpin and Ames, Eds., 1980.

attributed to bigfoot. Using sophisticated equipment and data analysis methods, they attempted to define the acoustic parameters of an utterance, leading to complex diagrams and commentaries, peppered with long mathematical formulae. According to them, the recording made at an altitude of 8,500 feet (2,800 m) in the High Sierras of California (after the first snowfall, in the night of October 21, 1972) suggested the presence of a number of unclassified creatures. One of them, at least, had to exceed the stature of an average human being.

Anthropologist George W. Gill of the University of Wyoming examined 200 reports of footprints and heights estimates, gathered over 94 years from Oregon to British Columbia, in an attempt to describe the distribution of the sasquatch population. More learned diagrams were presented, leading to the following conclusion: Either the greatest hoax in the history of anthropology has gone unmasked for centuries, or the largest nonhuman primate has survived in North America without being detected by modern science. While neither of these options seemed admissible, the author believed that one of them had to be true.

Two other academic researchers examined five specimens of fecal matter and three of hair attributed to Sasquatches. Two of the fecal samples and one of the hair samples belonged to known animals but the origin of the other specimens remained mysterious. The investigators regretted that an insufficient number of samples were available to arrive at a definite conclusion as to the existence of the sasquatch—which still remains an open question.

The Vancouver symposium clearly drew on a wide range of expertise. Many of the published reports require serious attention, although those relating to the cultural aspects of the sasquatch phenomenon are easier to access. They include reflections on the world of elves, medieval monsters, the concept of the wild man, and American Indian traditions. These most interesting contributions make up about four-fifths of the proceedings of the symposium.

As to Dmitri Bayanov, he deplored that such an emphasis had been given to specialists in cultural anthropology, mythology and folklore, especially since his contribution, as well as those of Igor Bourtsev and Marie-Jeanne Koffmann,[2] were omitted from the proceedings published by the convenors—perhaps not such an unusual failure of standard academic ethics. The vaunted reputation for openness of North American universities ended up somewhat tarnished by Marjorie Halpin's decision, justified to the Russian team as follows: "In order to give the problem the greatest possible chance to be taken seriously by the academic community, we cannot risk weakening the book by presenting an argument that can be dismissed by the scientists. I trust that you will understand that the issue is too important to be compromised."[3] Grover Krantz' s take on the symposium is also of interest:

> The conference was interesting. Perhaps more good than bad. Many attending were what I call " true believers" who have faith in the subject, without knowledge...Fortunately they did not disrupt things. Another group was the folklorists, reputable anthropologists, who just wanted to give a paper at a scientific meeting. Many of them did not care if Sasquatch, or the like, were actual creatures...Halpin said a book of all conference papers should be out in less than a year...Some discussions and questions answers were not prepared but they were recorded; I hope they will be included"[4]

Krantz wrote in moderate terms. One might imagine that his tone changed after he discovered that his own contribution was also omitted from the proceedings.

Overall, in spite of the criticism it generated, the Vancouver symposium succeeded in stimulating interest and curiosity.

---

2. Contributions by American presenters Carleton Coon, a respected senior anthropologist, Grover Krantz and Vladimir Markotic, were also omitted. Most of these were subsequently published by Markotic in *The Sasquatch and other Unknown Hominoids*, Western Publishers, Calgary, 1984.
3. Letter from Marjorie Halpin (June 29, 1979) to Dmitri Bayanov and Igor Bourtsev; cited by Bayanov in *America's Bigfoot, Fact Not Fiction*, p. 168.
4. Krantz's letter published in Bayanov's *America' s Bigfoot: Fact Not Fiction*, p. 161.

Chapter 34

The Father of Cryptozoology

The Aardvark, *(Orycteropus afer* [*"digging foot")* is a medium-sized, burrowing, nocturnal mammal native to Africa. It is sometimes called antbear, anteater, Cape anteater (after the Cape of Good Hope), earth hog or earth pig. The name comes from the Afrikaans/Dutch for "earth pig" (*aarde* = earth, *varken* = pig), because early settlers from Europe thought it resembled a pig. However, the aardvark is not closely related to the pig; rather, it is the sole recent representative of the obscure mammalian order Tubulidentata, in which it is usually considered to form a single variable species of the genus *Orycteropus* (Wikipedia).

In the 1950s growing public curiosity about the mysteries of nature—particularly the animal world—was fueled by a variety of publications, articles, photo documentaries and movies. A prime example is Romain Gary' s novel *Les Racines du Ciel* (winner of the Prix Goncourt in 1956), wherein the hero devotes his life to saving African elephants. The American director John Huston adapted it to the screen in 1958 *(The Roots of Heaven).*

Incidentally, it is in that novel that the word "ecology" makes one of its earliest appearances in mainstream literature, having hitherto remained a rarely used term, except among biology specialists. That word "ecology" was to greatly grow in significance in the years to come.

A new word that also made its appearance at that time, was "cryptozoology" defined by its originator, Bernard Heuvelmans as follows: "I crafted it from the Greek roots *kryptos* = hidden, *zoon* = animal, and *logos* = word: the science of hidden animals."

Could anything serious be expected from a man who had written a doctoral thesis in zoology on the dentition of the aardvark, the Boers' " earth pig?" The scientific establishment had a good chuckle—I heard it myself—"Ha! Ha! He began his career by studying the teeth of a toothless animal!"

Moreover, Bernard' s career had not followed the normal well-trodden scientific path. Author of excellent popular books on

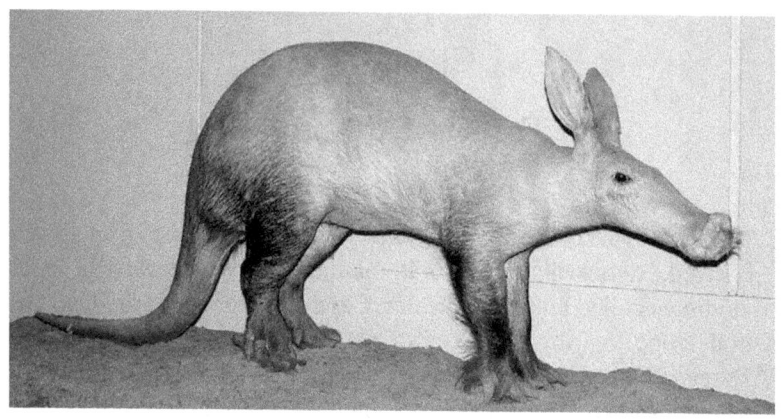

The aardvark *(Orycteropus afer)*.

astronomy and physics, he was also a jazz singer who had written a history of Black American music, *De la bamboula au be-bop*, and contributed to a wide variety of magazines. I was enchanted by his jazz expertise and really came to appreciate his sense of humour when I showed him a collection of translations and commentaries by Marguerite Yourcenar *(Blues and Gospels,* 1984*)*, a writer whom we both held in high esteem. Bernard immediately noticed the bizarre contrast between the slangy American original and the sophisticated Yourcenar translation. He was elegantly at ease in English, both written and spoken, and could easily maneuver between the Queen's English and street talk. Not surprisingly, since he had a Belgian father and a Dutch mother, he was also fluent in Dutch and in German.

A restless child, Bernard loved adventure stories. Three novels were to make a lasting impression on him: *Twenty Thousand Leagues Under the Sea* by Jules Verne (1828–1905)—recall the encounter between the Nautilus and the giant squid; *Lost World* by Conan Doyle (1839–1930) where prehistoric animals are imagined to have survived on a South American plateau; *Les Dieux Rouges (The Red Gods)* by Jean d' Esme (1894–1966), which describes the twentieth-century discovery of a tribe of ape-men in Indochina. One can't emphasize enough the degree to which such fantastic fiction can contribute to the blossoming of a scientific vocation.

Bernard also authored writings that stimulated the imagination and creative thinking. The most famous of his books is probably his first: *Sur la piste des betes ignorées* (in 2 volumes, Plon, 1955), recently translated and updated, with an introduction by Gerald Durrell (*On the Track of Unknown Animals,* Kegan Paul International, London, 1995). This volume includes many original drawings by Alika Lindbergh who shared the author's life from 1950 to 1960. In her words:

> Seen from this angle, science is a fantastic adventure. How could one not be enthralled by the wild man of the Caucasus—which I painted many times, with his pensive gaze typical of the Great Apes—or not dream of the dragon said to haunt the rivers and swamps of the Congo and thought to be a surviving dinosaur? What fairy wouldn't like to live among those dragons and fabulous beasts? What woman wouldn't be overwhelmed to discover that they are indeed part of tangible reality? Sharing the life of a cryptozoologist was to live in Conan Doyle's *Lost World.* Some day, Bernard would receive in the mail (from faraway Nepal, for example), a tuft of stiff hair; the next day would bring in excrement from the *Abominable Snowman,* for analysis by the scientists of the Belgian Royal Institute of Natural Sciences. Such dispatches from every corner of the world were apt to excite the imagination: be it a plaster cast of the enormous foot of Bigfoot, or the photo of a giant python, taken from a helicopter, revealing a gigantic reptile, at least a dozen meters [40 feet] long, slithering between the trees of the savannah.[1]

Thanks to his international network of correspondents, Bernard had at his disposal myriad sources of information which he scrutinized and carefully classified. His files bore the names of prestigious informants: Homer, Aristotle, Pliny, Ambroise Paré, Georges Cuvier, as well as contemporary contacts whom he met at congresses or on trips abroad. He paid close attention to the accounts of travelers and mariners; he sifted through the descriptions of " monsters" such as the kraken, or the giant octopus, given by major authors. In

---

1. Alika Lindbergh in *Testament d' une Fée,* p. 61 Editions e/dite, 2000.

this regard, Bernard took advantage of the zoological faux pas of Victor Hugo (*The Toilers of the Sea*, 1866) and Jules Verne (*Twenty Thousand Leagues under the Sea*, 1870) to discreetly identify the limitations of the imaginary. Hugo's octopus is truly disgusting:

> It has a single orifice, at the center of its tentacles. Is this unique hole its mouth? Its anus? It is both! The same opening serves both functions. The entrance is also the exit."

Bernard remarks:

> That poor monster must have truly terrible breath! Actually, among all Molluscs, and particularly among nonliterary Cephalopods, the anal orifice is always clearly distinct from the mouth.[2]

Heuvelmans' erudition and his lively writing proved extremely popular; translations of his works—particularly into English—allowed his writings to be known far beyond France. The British, keen observers of flora and fauna, embraced his works. The Americans, proud of their wilderness and vast unspoiled spaces, paid special attention to Bernard's books. In some of the most unspoiled states people are proud of their own personal wildlife libraries. The citizens of Alaska, Minnesota or Washington State, for example, often feel the need to deepen their knowledge of the natural world which surrounds them. I have never seen as many writings on animals in my own country.

---

2. Bernard Heuvelmans, *The Kraken and the Colossal Octopus*, (Kegan Paul, London, 2006), p. 36.

Chapter 35

The Impact of the ISC

Interest in undiscovered animals was greatly stimulated by the creation in 1981 of the International Society of Cryptozoology (ISC), presided by Bernard Heuvelmans. The international composition of its board of directors reflected the breadth of its views.[1] Distinguished honorary members were created at its first meeting: Marjorie Courtenay-Latimer, Marie-Jeanne Koffmann, Théodore Monod, John Napier, André Capart, David James, Ingo Krumbiegel and Sir Peter Scott.

The ISC published a quarterly illustrated newsletter, which touched on a wide range of topics. There was an in-depth discussion on the authenticity of photos of Nessie, the Loch Ness creature, taken in 1972 with an underwater camera by the Academy of Applied Sciences of Boston, USA. Another hot topic was the potential survival of the thylacine, or Tasmanian tiger, wiped out by sheep ranchers, the last specimen of which died at the Hobart zoo, Tasmania, in 1936. A solidly documented article in *The ISC Newsletter* (winter 1985) raised a ray of hope: recent eyewitness reports suggested that the strange, zebra-striped, marsupial carnivore might have survived after all.

Excellent interviews with René Dahinden on bigfoot, or with Marie-Jeanne Koffmann on the the wild men of the Caucasus, captivated the readers' interest. Dahinden's spicy expletives were reproduced verbatim, setting a sharp contrast with Koffmann's more refined language.

The newsletter heralded the discovery of a new species. In

---

1. Founding ISC directors were: Heuvelmans (president), Roy Mackal (USA, vice-president), Richard Greenwell (USA, secretary), Dmitri Bayanov (USSR), Eric Buffetaut (France), Joseph Gennaro (USA), Philippe Janvier (France), Grover Krantz (USA), Paul LeBlond (Canada), Nikolai Spassov (Bulgaria), Phillip Tobias (South Africa), Leigh van Valen (USA), Forest Wood (USA), Zhou Guoxing (China), George Zug (USA).

1985, German zoologist Wolfgang Bohme was watching a television documentary on Yemen. A short sequence, showing a varan lizard climbing a tree, attracted his attention. He obtained a copy of the show. The next step was to send an expedition to the Tihamat desert, the "hot land," to obtain live animals. A Swiss expedition led by Zurich herpetologist Beat Schatti captured eight four-foot-long (1.20 m) varans. Two specimens were sent to Dr. Bohme, in Bonn. Luck, a sharp eye and the sleuthing talents of W. Bohme and his Swiss colleagues led them to add a new species to the 30 known members of the varan family, the largest being the Komodo dragon of Indonesia, which is up to 10 feet (3.0 m) long.

During a meeting in Bonn in 1989, ISC secretary Richard Greenwell commented that this was the first time he had touched a live specimen of a recently discovered species, actually a species that had not yet received an official name.

The noted ISC member, Sir Peter Scott, died in 1989, shortly before his eightieth birthday. The ISC Newsletter published a lengthy biographical obituary, noting that from 1958 on, Sir Peter had been interested in Nessie, while diving in the loch, or flying over it in a glider. Scott thought that the animal might be a surviving plesiosaur, a marine reptile of the Jurassic period (130–150 million years ago), 10 to 16 feet (3–5 m) long, with a long neck mounted on a large body propelled by four large flippers.

> We were not concerned with the fact that no fossil plesiosaur had been found after the Jurassic. This was also the case for other animals, like the coelecanth and other surviving fossils.[2]

In 1975, Scott coauthored an article in *Nature*[3] with Robert Rines, president of the Boston Academy of Applied Sciences, in which they ventured to give Nessie a scientific name: *Nessiteras rhombopteryx*, which means the "Nessie beast with a diamond-shaped flipper." Lacking a specimen one may, according to the rules

---

2. Quoted by J.J. Barloy, "Serpent de mer...," p. 65.
3. Scott, P. and R.Rines, 1975. "Naming the Loch Ness Monster," *Nature*, pp. 258, 466–468.

of international zoological nomenclature, use a photo or an illustration to describe and name a new species. However, the illustration provided by Scott and Rines was far from convincing to many skeptics. It was pointed out that the letters of the new species were also an anagram for "Monster hoax by Sir Peter S." Sir Peter's reputation suffered of course, and one should admire his courage in facing such jibes.

Peter Scott was the son of Captain Robert F. Scott, who perished from the cold on his return from the South Pole in 1912. Peter studied in Cambridge, was a national gliding champion, and won a bronze medal in rowing at the Berlin Olympics of 1936. He became a renowned animal painter. He fought with the Royal Navy during the World War II, was decorated, and returned to his artistic endeavors after the war. His favorite pastime was duck hunting. However, one day he wounded a goose which agonized for hours in an unreachable swamp. It was then that Scott decided to hang up his gun for good. Some hunters are thus struck by remorse, even among the most prominent: George Adamson, for example, or Peter Byrne come to mind.

Scott became one of the leaders of the movement that in 1961 led to the creation of the World Wildlife Fund. He was its president and then its vice-president for 12 years and designed its famous logo—the giant panda within a circle. He was knighted by Queen Elizabeth II in 1973, and was thus the first to be rewarded for his work to protect nature. His life and work well deserved the homage that he received from the ISC.

The ISC convened congresses in the USA and in Europe, sometimes jointly with other institutions, like the British Folklore Society. The ISC newsletter presented summaries of the contributions representing the interests of the various presenters.

It is thus that in 1985, in Brighton, at a conference organized by the University of Sussex, Polish linguist and philosopher Piotr Klafkowski, met with well-deserved success. Piotr had left the stifling social climate of his country where nobody was interested in his expertise. He knew an impressive number of rare Indo-European, Slavic, and even Finno-ougric languages. He scrutinized ancient texts so as to discover the structures upon which these societies had been based. His approach was comparable to that of

Georges Dumézil (1889–1986), a reputed linguist and specialist of comparative mythologies.[4]

Piotr had found refuge in Oslo, Norway, where he was living modestly, pursuing his linguistic interests, apparently meaningless since they were of no use either to the East or the West. By what chance did the ISC, through its secretary Richard Greenwell, discover this extraordinary solitary researcher? I don't know. His glory was quite ephemeral. I twice enquired with Greenwell about Piotr's whereabouts—in vain. The ISC had lost track of him. One had to wait a few years to find his work widely referenced on the Internet.

Piotr Klafkowski

The ISC Newsletter (Fall 1985) described Klafkowski's presentation in Brighton as follows:

> The final talk in the morning session was by Piotr Klafkowski, a Polish linguist currently residing in Oslo, Norway, who specializes in obscure, unwritten and vanishing languages. Dr. Klafkowski spoke on "The Case for a Linguistic Component in Cryptozoology," during which he described how the extinction of languages often occurs— in unison with the destruction of habitat and species—before they can be described and studied. He stated that cryptozoological data can be gathered from vanishing languages, both in the field and in the library, and that linguistic analysis of such data can provide clues for planning cryptozoological field work. Dr. Klafkowski also suggested the establishment of a "linguistic information bank" within cryptozoology.

---

4. Georges Dumézil was a French comparative philologist best known for his analysis of sovereignty and power in Proto-Indo-European religion and society. He is considered one of the major contributors to mythography, in particular for his creation of the trifunctional hypothesis of social class. (Wikipedia)

Somewhat incidentally, Piotr had discovered that the languages he was studying dealt with mythological themes filled with creatures of interest to cryptozoologists. He spoke in English, with great precision. His initial awkwardness and unfashionable clothes were soon forgotten. The audience was moved by his passionate approach to his subject and by the depth of his erudition—his presentation raised a standing ovation. For too brief a moment, moved by the speaker's intelligence, courage and dedication, we had been buoyed by the originality of his thinking.

Such an originality of thought, found in the works of Bernard Heuvelmans, as well as in the publications of the ISC, was a breath of fresh air for many readers: those from beyond the Iron Curtain, under the thumb of intellectual dictatorship, as well as those in the "free" countries, still submitted to the authority of a conservative establishment.

Chapter 36
About Some Other Unusual Folk and Creatures

Letters to the ISC Newsletter gave rise to lively and fruitful exchanges. The newsletter also featured numerous works of zoological and cryptozoological interest. I had the pleasure of contributing to its review of books.[1] Bernard Heuvelmans also asked me to translate some articles from French to English to send to the editor, Richard Greenwell. The discovery of new species was deemed of particular interest to cryptozoologists and often reported in the newsletter. As the following examples illustrate, new species come to light in varied and unexpected ways.

In 1987, I noticed an article describing the unexpected discovery of a new species of gecko, the largest known to date.[2] It all started in July 1984, during the war between Iran and Iraq. An Iranian stretcher-bearer ran after an animal fleeing the nearby explosion of an artillery shell. Thanks to Corporal Ali Reza Ensaf and his presence of mind in difficult circumstances, a specimen of this rare gecko, baptized *Eublepharis ensafi,* is preserved today in the Faculty of Science of Teheran.

A spectacular discovery was summarized in the 1989 issue of *Cryptozoology,* the scientific journal of the ISC: "Recollections on the discovery of the Coelacanth Latimeria Chalumnae Smith," wherein the author and discoverer, Marjorie Courtenay-Latimer, reminisced over the circumstances surrounding the "zoological discovery of the century," 50 years earlier.

Marjorie Latimer and the coelacanth.

---

1. See Appendix 3.
2. *L'Univers Vivant,* March 1987, pp. 95–97.

In 1931, Miss Latimer began her career as a curator in the small East London museum on the southeast coast of South Africa. She was a friend of Hendrick Goosen, master of the trawler *Nerine*. On December 22, 1938, following a phone call, Miss Latimer took a cab down to the harbor; a number of specimens had been put aside for her consideration—sharks, rays, sponges, corals. Among them was the most beautiful fish she had ever seen. The taxi driver was at first reluctant to carry it—the dead fish was going to stink up his trunk. Back at the museum, she consulted her reference books. The animal was apparently a member of the ganoid superfamily (which includes the sturgeon), most similar to the extinct Coelecanthidae family. But, she thought, "A ganoid fish is a fossil fish (mostly). This is a live fish! It cannot be a fossil fish! In any case, I knew that something had to be done to save this specimen; but how?"[3]

The rest of the story is part of the legend. Miss Latimer had a whole series of obstacles to overcome. First of all, disbelief. For museum chairman Dr. Bruce-Bayes, it was just a "big rock cod," a common enough fish found on rocky coasts. Although its flesh is said to be mediocre-tasting, it is often found for sale as "rock-fish" on Jersey Island fish stalls. Its dark green color, spotted with red, yellow and blue, give it an ancient appearance, which puts off customers. This fascinating creature is reminiscent of the coelacanth, a full-size model of which is on display at the Croisic aquarium.[4]

The reader will understand why I find myself moved by the series of difficulties related to the recognition of the coelacanth. Miss Latimer had to face the refusal of the director of the hospital morgue and then that of the director of the cold storage plant.

Her mother agreed to rip some sheets into strips to help an aged taxidermist, Robert Center, who consented to assist the young woman. The idea was to ration the small supply of formalin available to soak the strips with which to wrap the fish.

Miss Latimer attempted to phone Professor J.B.L. Smith at Rhodes University College, an expert in ichtyology. Smith was

---

3. *Cryptozoology,* Vol. 8, p. 6.
4. Le Croisic is a small fishing port on the coast of Brittany. Its aquarium is a major tourist attraction. Stuffed coelacanths may also be seen in a number of other aquaria around the world.

away and did not respond until January 9, 1939. Over the days, the carcass disintegrated. However, the taxidermist had done his best and on February 16, Dr. Smith set his eyes on the survivor from the remote past. Latimer recalled her meeting with Smith.

> How old is it?, I asked. Smith counted the rings on the scales and said that it was about 33 years old, but its primitive origins were more like 70 million years old. My mind boggled. So I had been on the right track. It was a living fossil.
>
> On March 29, 1939, Smith phoned to say that he had named the fish *Latimeria chalumnae*, after me and the Chalumna River, in the mouth of which it had been caught, at 40 fathoms."[5]

J.L.B. Smith later wrote:

> Now, in the shock of the discovery, scientists all over the world were busy considering the problem of how such a large and curious-looking creature had managed to escape notice all this time.[6]

Smith's book describes the ins-and-outs of the discovery of the living fossil, as well as the problems arising in its formal classification because of the skepticism of some scientists.

Smith's work appeared to me as fascinating as the adventure stories of Jack London or Joseph Kessel. It also led to a new perspective on biological evolution. If the coelacanth could have persisted unchanged for hundred of millions of years, it followed that some species evolve only little, if at all. Instead of changing slowly and gradually, species might change rapidly under special conditions. This is the idea that American paleontologists Nils Eldridge and Steven Jay Gould put forward in 1972 as "punctuated equilibri-

---

5. Cryptozoology, Vol. 8, p. 7,8.
6. J.B.L. Smith, *Old Four Legs: The Story of the Coelacanth*, p. 59.

um."[7] For a new species to appear, a relatively short time might be required, perhaps as few as 5,000 years.

How would a new species appear? All that's required is that some group of individuals become separated from the rest of the species. Genetic mutations take place all the time; when they are not shared within the whole population, separate groups become sufficiently different and speciation takes place. The ISC's logo, the okapi, illustrates the phenomenon.

According to Bernard Heuvelmans, the giraffe and the okapi both belong to a "large group of early giraffids widespread in Africa and Asia in the Miocene and especially in the Pliocene." Heuvelmans goes on to explain how the okapi was identified, in 1900.

There were some early clues, found by Stanley, the journalist-explorer who discovered that the Wambutti [pygmies of the Congo's Ituri forest] knew a donkey which they called atti. They said they sometimes caught them in pits and that they ate leaves.

Sir Harry Johnston, then governor of Uganda, was visiting the Congo. His curiosity was piqued by Stanley's words. In 1899, Sir Harry:

> saved a party of pygmies from being carried off by a German showman to be exhibited as curiosities at the 1900 Paris Exhibition. They were his guests in Uganda for several months until he could take them back to their native forests. Meanwhile, they became firm friends.[8]

Once in the Congo Sir Harry continued his search for the okapi. On the strength of the pygmies' description and a few anatomical relics (skull, pieces of skin) it was possible to reconstruct the beast. In Heuvelmans' words: "It was a strange beast, reminiscent of those

---

7. Punctuated equilibrium is a theory of evolutionary biology which states that most sexually reproducing populations experience little change for most of their geological history, and that when phenotypic evolution does occur, it is localized in rare, rapid events of branching speciation. (Wikipedia)

8. The story of the discovery of the okapi is described in detail in Heuvelmans' *On the Track of Unknown Animals*, p. 44.

mythological monsters created by patching together parts from many different creatures."

In profile it looked somewhat like a horse or an antelope; its tongue was that of an anteater, its ears similar to a donkey's, and the stripes on its hindquarter reminiscent of those of the zebra. One can imagine that as the savanna shrank in area (it has completely disappeared from the Sahara and from Egypt), the giraffids, in competition with the antelopes, became used to eating leaves. Some took refuge in the forest and became today's tan and striped okapis—*Okapia johnstoni*. The giraffe and the okapi had enough time for the phenomenon of speciation to proceed gradually, over a few million years.

Speciation however sometimes occurs much faster. As an example, take the sockeye salmon of Lake Washington, in Seattle, monitored since 1930. This large lake, one of the many within the city area, is linked to the Pacific Ocean via the smaller Lake Union and a series of locks and fish ladders allowing salmon to swim upstream into fresh water all the way to the Cedar River, which flows into Lake Washington.

These sockeye have split into two distinct groups over the past 60 years; some live near the shore of the lake, other in the Cedar River. The near-shore males have grown by 10 percent and have become more streamlined compared to those living in the stream. The in-stream females on the other hand have grown even larger, gaining in strength to dig their redds in the fast-flowing river. The differences between the two populations have given rise to reproductive isolation: they can no longer interbreed, even though they are not separated geographically. Still, how can one explain how these two groups have drifted apart?

Bernard Heuvelmans would probably suggest that one should wait a few years for confirmation of the small genetic differences. He might add that these salmon were already different at the beginning, possessing slightly different chromosomic baggage. What seems difficult to explain is the original divergence in the behavior of the sockeye salmon, the first steps toward speciation. Clearly, my background does not permit me to attempt an answer. However, various reputed scientists have suggested, in plain language, hypotheses and tentative explanations.

Ian Tattersall, for example, an expert in the evolution of humans and primates, sheds interesting light on the fascinating questions raised by cryptozoology.[9] A number of French authors of similar standing (Yves Coppens, Pascal Picq, Henry de Lumley) have contributed to the discussion, extending and criticizing the work of cryptozoologists.

One of the strangest known animals is undoubtedly the "water mole" (platypus) discovered in 1797 in Australia. In 1801 the German zoologist Blumenback named it *Ornithorhinchus paradoxus,* "the paradoxical bird-beaked creature." For years, European zoologists like Cuvier, Geoffroy Saint-Hilaire, Blumenbach, Meckel and Owen argued about the nature of this animal with a duck's beak, which lays eggs but also nurses its young.

By 1884, the Australian Caldwell and the German Haacke demonstrated that the anatomically paradoxical beast was indeed an egglayer. Citing Baron Cuvier's rash pronouncement in 1812: "There is little hope of discovering new species of great quadrupeds,"[10] Heuvelmans showed by a variety of examples how wrong the father of paleontology had been, repeatedly emphasizing: "One thing was by now undeniable: impossible creatures can exist."

At this point, I cannot resist pointing out some of the spectacular discoveries of large hitherto unsuspected animals described by Heuvelmans.

You have perhaps seen in a zoo, if not in its natural habitat in the steppes of Asia, a kind of shaggy wild horse of an obviously primitive type. With its large head, short legs and thick winter coat it's easy to mistake it, at a distance, for a donkey. This is the only surviving wild horse, discovered in 1881 by a Russian explorer, cavalry officer Nikolai Mikhailovich Przewalski, who indeed mistook it for a wild ass. Zoologist M. Poliakoff took pains to convince him that the animal was a horse, which he named *Equus przewalski.*

Furthermore, one should not forget Father Armand David, a French missionary who between 1865 and 1869 identified three large mammals in eastern Asia.

---

9. Ian Tattersall is currently curator in the Department of Anthropology of the American Museum of Natural History in New York City and author of many books on human evolution, for example, *Becoming Human,* 1998, Harcourt Brace, NY.

10. Cuvier, as quoted by Heuvelmans, *On the Track of Unknown Animals,* p. 18.

The first one was a deer which he helped rescue from extinction. Today, one can see it in some zoos as "Father David's Deer." The adventurous circumstances that led to the discovery of the animal in an imperial park illustrate how success often requires that scientific knowledge and intuition be supported by courage, stealth and persuasiveness.[11]

The second animal, represented by Chinese artists on vases or on silks, looked like a devil with red coat and upturned nose. To Father David's eyes, it was a monkey, an idea that would probably have provoked laughter among most naturalists of the time. But he was right: there does exist a snow monkey, living in the Himalayas, up to an altitude of 1,300 feet (400 m). It has a thick coat and an upturned nose and is commonly known as the golden snub-nosed monkey. In 1870, the French zoologist Alphonse Milne-Edwards examined a pelt and named the animal *Rhinopithecus roxellanae*.

Incidentally, there exist in the Himalayas two species of monkeys which, standing on their hind legs, reach 48–55 inches (120–140 cm). Besides the snub-nosed monkey, there is also the Himalayan Langur *(Semnopithecus entellus)*, distinguished by its sturdy build. It is a rare animal and some have claimed that its footprints have been taken for those of the snowman. However no one has managed to explain the distance between successive footprints of the latter.

Father David was later to add the third rather stunning discovery to his report card. On March 11, 1869, while a guest at the house of a rich Szechwanese land owner, he saw the fur of an animal that he had heard rumors about—a black and white bear living in bamboo forests.[12] He was immediately convinced of the animal's reality and strove to acquire a specimen. At his request, hunters set off to find one, and returned on March 23, 1869 with the body of a young giant panda, and a few days later with that of an adult.

Soon afterwards Father David acquired another, already known animal, the Hun-Ho or fire-fox. This small animal, about the size of a large house cat with a ringed tail, was dubbed the Himalayan rac-

---

11. Heuvelmans. *On the Track of Unknown Animals*, p. 29.
12. Heuvelmans adds that a Tang dynasty manuscript dated 621 AD mentions it; loc.cit. p. 31.

(Above) "Father David's bear," the giant panda *(Ailuropoda melanoleuca)*. Only About 2,000 to 3,000 are believed to be in the wild. However, the number appears to be on the rise.

(Right) The lesser panda or red panda *(Ailurus fulgens),* originally known as the *Hun-Ho* (fire fox). Fewer than 2,500 mature individuals are estimated to remain in the wild.

coon. Father David did not suggest a link between these two creatures, now known as the giant panda (or for some specialists, "Father David's bear") and the lesser panda respectively. After examining their teeth and their skeletons, Milne-Edwards removed them from the bear family and classified them with the *Procyonidae:* the raccoon family.

The giant panda soon morphed into a beloved icon. As early as 1936, Ruth Harkness brought back from China a live baby panda that became the darling of Chicago's Brooksfield zoo. A comic-strip character, Andy Panda, soon followed and toy makers made a roaring business selling stuffed panda bears.

Meanwhile, caring little for the controversies about their classification as bears, procyonids (raccoons) or, according to Dr Frechkop, ailurids (a group intermediate between bears and cats), the pandas go on with their lives, the lesser ones napping on a branch, their giant cousins sitting on their rumps munching on bamboo shoots.

As yet another example of the promise of the cryptozoological method, I would like to bring up another fascinating animal, a beast that sounds at first like a creature of fantasy—a deadly, venomous dragon. We shall then return to our favorite subject, the wild men, both large and small, where we shall find a discovery at least as and perhaps even more surprising than Father David's deer or pandas—small fossils of the genus *Homo*.

It was thus in 1912, also in the Orient, on Komodo Island in the Sunda archipelago, between Sumbawa and Flores islands, that after an emergency landing a pilot discovered a giant reptile, reaching up to 13 feet (4 m) in length. "It was soon found that the dragon *(Varanus komodensis)* was not confined to Komodo Island, but also lived on the small islands of Ritja and Padar, at the eastern end of Flores Island," commented Bernard Heuvelmans.[13]

This is the region that recently (December 26, 2004) experienced a major underwater earthquake, northwest of Sumatra. The giant wave that followed flooded the coasts of 11 countries bordering the Indian Ocean, causing at least 200,000 casualties. The feeling of sadness provoked by this catastrophe is amplified by the

---

13. Heuvelmans, loc. cit. p. 52.

The fearful Komodo dragon *(Varanus Komodoensis)*. It has been recently discovered that female dragons are capable of parthenogenesis, in which viable eggs are laid without fertilizition by a male. There are between 4,000 and 5,000 dragons in the wild.

thought that, thanks to Bernard Heuvelmans, Indonesia has been found to be the home of marvelous creatures, related perhaps to the last hominoid survivor of the Pongidae family, the orangutan (literally 'man of the forest') of Sumatra and Borneo. After studying the development, the structure and the locomotion of the orangutan, Heuvelmans drew some analogies with the yeti whose evolution apparently proceeded in a similar fashion as the orangutan's, although tending more precociously toward a bipedal stance.

In a passage that remains etched in my memory, Heuvelmans put forward a bold hypothesis about the small Indonesian orang-pendek (small man in Malay) wondering if, "these hairy gnomes were midget *Pithecanthropes?*" He attempted to classify these small Indo-Malaysian creatures:

> they exhibit numerous traits which place them halfway between humans and anthropoids (like the gibbon and the orangutan), but distinguish them from Java Man *(Pithecanthropus erectus)* or Peking Man *(Sinanthropus),* which were of a stature similar to ours. [and he adds, a few lines later] recent paleontological discoveries have shown that there is considerable variability among *Pithecanthropes.* It would thus not be surprising to find that there were midget *Pithecanthropes*: in a sense, the Asiatic counterpart of the small Australian *Australopithecines.*[14]

---

14. Heuvelmans, loc. cit. p. 115.

Some of Heuvelmans' readers reacted with delight to the stories describing the orang-pendek; others boiled with impatience at such unrestrained imagination. Nevertheless, a corpus of historical and geographical details, numerous sketches clearly distinguishing the pendek from the Borneo proboscis monkey (or orang banda) and the Malaysian sun bear, the accuracy of the descriptions offered by the natives as well as by Dutch colonists, and the study of footprints, all add up to a fascinating and compelling file.

Half a century later, in October 2004, the discovery of a skeleton, dubbed a "hobbit" because of its small size, was to justify Heuvelmans' speculation. A scientific team led by eminent Australian paleontologist Richard Roberts (a professor at Wollongong University) discovered on Flores Island the nearly complete skeleton of a cave woman only 39 inches (1.0 m) tall. She lived 18,000 years ago, a member of a tribe that made tools, hunted the pygmy elephant and rubbed shoulders with the Komodo dragon.[15]

Flora, the 55-pound (25-kg) Flores cavewoman, was smaller than Lucy the 50-inch (127-cm) Australopithecine who lived 3.2 million years ago. Her small brain capacity (380 cm^3) was similar to that of the chimpanzee, but the shape of the *Homo floresiensis* skull relates it more closely to *Homo erectus*. As Pascal Picq noted, the discovery of such a diminutive species on Flores island shows that man, just as any other species isolated on an island, can evolve towards smaller size and brain capacity.

Their small size was no obstacle to the Flores islanders' survival from 95,000 years to about 13,000 years ago. Richard Roberts, like Heuvelmans, explored the local folklore, and uncovered stories which at first appear fantastic, but rich with so much detail that they must contain some kernel of truth. A village elder spoke of the customs of the *ebu gogo* [*ebu* = grandmother; *gogo* = she who eats anything]. According to local stories, the small people were about three feet tall (1.0 m) tall, had long hair, a round belly and a slightly awkward gait. These descriptions agree with what can be inferred from Flora's skeleton.

Researchers believe that the hobbits were wiped out by a vol-

---

15. For an account of the discovery of Flores Man and the controversies surrounding it, see Wikipedia, *Homo floresiensis*.

canic eruption which destroyed the eastern part of Flores Island 12,000 years ago. However, villagers maintain that the Ebu Gogo survived as late as the 19th century. Heuvelmans wrote that:

> The goegoeh which the Dutch colonists mentioned must be the same gugu described by the English diplomat William Marsden at the beginning of the 19th century."[16]

Roberts also wondered, as did Heuvelmans, whether Flores man or perhaps some close relative, might not still be alive today on some nearby island.

So, we now know that three species of man have coexisted: *Homo neanderthalensis, Homo floresiensis* and *Homo sapiens*. Discovery of the "hobbit" skeleton has added another branch to the evolutionary tree. There is still room for more twigs to graft onto it as new fossils are discovered. In an article honoring Heuvelmans' memory, Pierre Lagrange, a research scientist at the CNRS wrote:

Orang-pendek drawing by Alika Lindbergh, who was personally associated with my good friend Bernard Heuvelmans.

> The adventure of Flores Man has only just begun. Whether it be cryptozoology, long scorned as a mere pseudoscience but perhaps soon to be taken more seriously, or Javanese tales of strange forest demons, it is no longer a matter of a confrontation between "superstitious gullibility" and "scientific culture." There is a lot to learn from folk tales, probably as much as from paleoanthropological research.[17]

---

16. Heuvelmans, loc.cit. p 123.
17. Lagrange, Pierre, *Science et Avenir*, December 2004, pp.12–14.

The news of the recent discovery of a large, new species of monkey, the Arunachal macaque *(Macaca munzala)* would have pleased Bernard Heuvelmans. The animal was discovered by an Indian ecologist in northeast India in the mountains of Arunachal Pradesh, where it lives between 5,200 and 12,000 feet (1,600–3,600 m). In a world where monkeys, especially the larger species, are disappearing (through destruction of their habitat and poaching) it is comforting to welcome this denizen of remote forests.

Still today, some dogmas survive. It is not so many years ago that experts were reluctant to admit the existence of surviving fossils, simply because they thought that archaic forms should have disappeared. Such survivors are witnesses of the past glory of formerly flourishing families. As to those fossils in which scientists often put so much confidence, they should be taken with some hesitation. As Heuvelmans points out, fossilization is a very chancy process, which explains the scarcity of the evidence available on the past history of men and apes.

Chapter 37
The Frozen Fossil

From his earliest childhood, Bernard Heuvelmans had the "most perfect rapport" with animals. Soon after his wedding, in 1939, he hastened to acquire a capuchin monkey. After the war, in 1950, he lived for ten years with his second wife, writer and painter Monique Watteau—today Alika Lindbergh—in a dark and narrow room in St. Germain des Prés, in Paris, with only cold running water. The couple also looked after some "exotic souls," described by Alika Lindbergh: "Boulimic, the sweet and overactive little capuchin monkey and later Belinda, the tender hairy monkey, were even more deprived of sunlight than we were."[1]

Over the years, the couple brought their furry companions for summer holidays on the Ile du Levant, where they enjoyed a feast of strawberries and cicadas.[2]

After the couple split up—Monique-Alika now becoming Mrs. Lindbergh upon marrying Scott, the son of Charles Lindbergh the aviator—they kept on friendly terms, sharing their love of animals until Bernard's death in 2001. Thus, in 1969, Scott and Alika received a letter from Bernard who had just purchased a howler monkey in Guatemala. The animal, locked up in a dark closet, was in serious danger. Bernard asked Scott and Alika to look after it, to which they agreed. Bernard managed to carry Ursula, as he named the monkey, hidden in a basket without anyone noticing on board his SABENA flight or at the customs in Brussels. Bernard was willing to take serious risks to protect animals. "He never betrayed the

---

1. Alika Lindbergh, *Testament d'une fee*, p 59.
2. The Ile du Levant is a small island off the Cote d'Azur shared by the French navy and a naturist club.

solemn pact which he made in his early childhood with every monkey of the world," adds Alika Lindbergh.[3]

Alika and Scott became very attached to the howlers, both red *(Alouatta seniculus)* and black *(Alouatta caraya)*. For 17 years, they raised these supposedly fragile wards on their property in the Dordogne. "We approached the matter as ethologists, first observing the affective, emotional and social needs of our wards, and then ensuring that those needs, be they psychological or physical, essential for the mental health of the animals be satisfied."[4]

The project involved the protection of an endangered species, with a record of rare survival in captivity, followed by their delicate reintroduction to their native habitat, the jungle of the Planalto of central Brazil. Scott and Alika succeeded in repatriating in Brazil three groups of monkeys, many of which had been born in the Dordogne. Bernard collaborated in this labor of patience, particularly when Scott and Alika were traveling overseas.

Alika Lindbergh's book, *When the Howlers will Fall Silent.*

The deep and instantaneous rapport that arises between some people and animals appears almost magical. This magic was evident when Bernard was in the company of a monkey, even more so with many monkeys. Thinking back to his years as a student in a Jesuit college, he confessed: "To St. Francis of Assisi, who humbly reached down to the level of the beast, he greatly preferred Tarzan, who saw himself as one of them and was disconsolate upon the tragic death of his monkey-mother, who had nursed and raised him."[5]

This attraction of Heuvelmans was combined with a major preoccupation—the problem of the origins of man. This is why he was quite overwhelmed when he found himself in the presence of an

---

3. Alika Lindbergh, *Quand les singes hurleurs se tairont*, p. 35.
4. Alika Lindbergh, *Testament d'une fée*, p 110.
5. Heuvelmans, *L'Homme de Néanderthal ...*, p. 15.

"escapee from prehistory," an encounter that happened in December 1968, as he was visiting his friend Ivan Sanderson, the writer and science reporter, in New York.

Sanderson suggested that Heuvelmans accompany him to Minnesota, nearly 2,000 miles (3,000 km) by car, to examine a frozen corpse on exhibit in local fairs. Holding back their skepticism, Heuvelmans and his friend were shocked when faced with the body preserved in a kind of refrigerated sarcophagus, visible behind a quadruple glass lid. The sarcophagus, held within a semitrailer, was being shown by Frank D. Hansen, a former captain in the US Air Force, and a veteran of Korea and Vietnam.

Summing up his observations Heuvelmans wrote:

> the being lying there under my eyes, in its frozen and lit-up sarcophagus, was characterized by a strange mixture of human and simian features.
>
> By the abundance and location of its body hair it was close to the great anthropoid apes, particularly the gorilla and the chimpanzee. His weakly opposing thumb was more similar to that of the lower primates, such as the New-World monkeys, the capuchins for example. The absence of lips and of a labio-nasal groove set him apart, together with all living primates, from human beings. Its limbs were quite similar in their proportions to those of some modern men but also, as one might easily forget, to those of some New-World monkeys. Its nose was clearly human in size, but its shape was strangely reminiscent of that of *Semnopithecus* or of the famous snub-nosed monkey, *Rhinopithecus.* Its vocal sac, if he really had one, would put him closer to the eastern great apes, the orangutan and the siamang gibbon.[6]

For three days, Heuvelmans and Sanderson examined, sketched and photographed the body. Their notes, photos and sketches were later used by Russian and American illustrators. The most painstaking renderings were undertaken by Alika Lindbergh under Heuvelmans' supervision.

---

6. Heuvelmans, loc. cit., p. 224.

Where did that six-foot-tall (1.8 m) creature with such convincing anatomical details come from? Even its wounds, probably lethal, looked authentic. How could one fake these "subtle defects: sores, scratches, local necroses, blood stains, etc.?"[7]

Frank Hansen offered at least three different explanations for the presence of the frozen creature. At first he said he had bought it from a businessman in Hong Kong. Then he said that some Californian mogul had seen it during a trip in the Orient and had asked Hansen to buy it so as to use it later as a fairground attraction. Later on, he described in a magazine article how he had shot it with a 8-mm caliber bullet while hunting in Minnesota.

For Heuvelmans, there was no doubt that the creature had indeed been killed by gunshot. From the evidence available and from Hansen's more extensive description of the wounds he reconstructed the story as follows: captured after being wounded by a low-caliber bullet in his right leg, the creature had remained lame and perhaps gangrenous. Angered by pain, it had tried to escape; hit on the forearm by his captors, they had then shot it, fearing its desperate strength; the bullet had entered its right eye and "the shock wave had pushed the other eye out of its orbit and created a large crater at the back of the skull, causing immediate death."[8]

The face of *Homo pongoides*, as drawn by Alika Lindbergh from photos and sketches by Bernard Heuvelmans.

Was the hairy man our contemporary? Which race did it belong to? While I would love to quote more of Heuvelmans' speculations,

---

7. Heuvelmans, loc. cit., p. 233.
8. Heuvelmans, loc. cit., p. 227.

which I found as captivating as his story of the discovery of the frozen man, we'll have to be content with a summary.

The time has come to use the name given by Sanderson and Heuvelmans to their discovery. As it was clearly a hominid, they decided to call it *Homo*. Heuvelmans suggested the species name "pongoid," meaning "looking like an anthropoid ape." Thus *Homo pongoides*, which Heuvelmans would have gladly made more specific as *Homo neanderthalensis pongoides*.

Heuvelmans thought that the pongoid man had been shot in Vietnam by the military on a hunting expedition. Returning home, Frank Hansen had decided to draw some benefit from the corpse of the relic Neanderthalian. Stored in a plastic body bag and enclosed in a sealed coffin, just like the corpse of a soldier killed in combat, pongoid man could have readily been transported back to some US Air Force base. With the right contacts, Hansen just had to pick him up there.

According to Heuvelmans, Hansen's later claim to have shot the creature in Minnesota would have been based on his fear of being accused of smuggling and illegal use of military equipment. There was no doubt in Heuvelmans' mind that the victim had been, just a few months earlier, "a living human fossil." Its combination of human and simian characteristics tended to classify it in one of the three categories considered intermediate: *Australopithecus, Pithecanthropus* or Neanderthal.

One should remember how vague these intermediate categories are. There might be more: Flores Man has just recently joined the list.

Hansen's hairy man was closest to the type that most recently vanished, Neanderthal man, because of some specific anatomical traits, described by Heuvelmans as follows:

> A strikingly wide face, pronounced eyebrows and sloping forehead, a massive chinless jaw, the head sunken into the shoulders, curved collar bones, the barrel chest, thick and short forearms, similarly short lower legs, enormous hands, surprisingly large feet, and finally, splayed and crooked toes.[9]

---

9. Heuvelmans, loc. cit., p. 254.

In a footnote, Heuvelmans adds:

> The in-depth examination of the photos of the external anatomy of the specimen, on which I spent two years, not only confirmed my early prognosis, it also revealed specific and much more subtle traits, linked to its peculiar snub-nose, the length of its thumb as well as its being barely opposable.[10]

Bernard Heuvelmans' photos, Sanderson's sketches, the renderings done by different illustrators (one American, the other Russian), without forgetting Alika Lindbergh's magnificent reconstitutions, add up to a most compelling graphical corpus. Superposition of the feet, hands and faces of the specimen with those of a Neanderthal and of a *Homo sapiens* confirm Heuvelmans' hypothesis by bringing out the similarities as well as the differences.

Eminent American anthropologist Carlton Coon was of the opinion that the frozen specimen was not a hominid, but a human being with a number of unexpected anatomical traits.[11] However, some of these traits are precisely those that would be found in a creature adapted to survive cold conditions, the classic Neanderthals, for example. They would look a lot like hardy mountain dwellers, to the point that Coon observed that even today one could find in the Abruzzes, the Alps and in Bavaria, people built more or less like Neanderthals.

What was then the result of this adaptation to cold climates? Heuvelmans and Porchnev think:

> through the ages, as Homo sapiens more or less retained a stable anatomy while developing its brain and its psychological and social abilities, Neanderthals, perhaps because of

---

10. ibid.
11. Prof. Coon's expertise was also sought in another mystery—that of the ethnic origin of the man whose outline appears on the Turin shroud. In both cases, the victim suffered a cruel death, following a series of blows and wounds.

their isolation in some regions of Europe and central Asia, and under the influence of the harsh climate of the Ice Ages, began to evolve and specialize more rapidly, its anatomical features becoming more and more characteristic.[12]

Professor Jean Piveteau considers Neanderthal man as a sterile branch of the human family tree, the victim of "a specialization that one should view as regressive."[13]

Bernard Heuvelmans, together with his friend Porchnev, had the audacity of claiming that some kind of hairy wild man had survived until today! Luckily, a few courageous experts dared support their views. For example, Professor André Capart, director of the Royal Belgian Institute for Natural Sciences, published Heuvelmans' preliminary analysis in the institute's Bulletin (February 10, 1969).

Clearly, the survival of *Homo pongoides* raised all kinds of questions. But don't we still find today hunter-gatherers coexisting with astronauts?

---

12. Heuvelmans, loc. cit. p. 395.
13. Heuvelmans, loc cit., p. 394.

Chapter 38

Lingering Questions

Bernard Heuvelmans raised a number of questions which, beyond a small circle of readers, did not stimulate many answers. Was it perhaps because they disturbed established scientific conceptions?

Heuvelmans suggested that Neanderthals devolved from a human cultural level to an "animal" state. He embraced Porchnev's hypothesis that, coincidentally with the thinning out of the great herbivore herds, stone tools had become rarer and even disappeared. Confronted with sapiens competitors, Neanderthals had retreated to inhospitable areas. Their decadence then became inevitable.

For Bernard Heuvelmans, *Homo pongoides* survivors would be the descendants of those devolved Neanderthals as well as of individual Neanderthals who, never having learned to make stone tools, had survived "precisely because of their animal nature."

Flight, dispersal, and a nocturnal lifestyle became synonymous with survival, but also brought cultural and social regression. These solitary beings forgot to honor their dead by digging and decorating graves as they had in earlier times. They perhaps continued to practice their rough devotion to the bear, an animal similar to them, sharing the same habitat of caves and forested hills.

Without speech, Neanderthals remained slaves to dumb imitation, incapable of communicating orally the details that enhance manual skills. Without speech, skills remain frozen.[1] As Heuvelmans loved to repeat: "Language, society and work are the three decisive factors of humanisation."

Such decisive factors as social organization and industry would have been of particular interest to the Soviet researchers gathered around Porchnev. In their day, and still today, the cultural level

---

1. Wild children, raised by wolves, never learn to speak or even to walk upright if they have been brought up alone by animals for the first years of their life.

reached by Neanderthals at the peak of their evolution remains surprising.

Furthermore, there is nothing simian about the Neanderthals' regression. Their stooped stance is related to their rock-climbing habits. As Heuvelmans points out: "To each his own specialization: anthropoid apes became forest trapeze artists, *Australopithecines* fearsome savanna hunters, *Paranthropes* powerful herbivore beasts, and Neanderthals expert mountain climbers."[2]

It is the least specialized, *Homo sapiens,* which thanks to his very lack of specialization managed to occupy a whole range of ecological niches. He then got rid of his relatives that filled niches he coveted.

Heuvelmans points out the inexhaustible "thirst for massacring its rivals and even mere bystanders" which is a "psychological trait dominating the lineage from which arose *Homo sapiens.*"[3] After decimating or even completely eliminating a number of populations, such as the Bushmen, the Australian Aborigines, the Tasmanians, the Maoris and the American Indians, the most advanced people still carry on their destructive work: oppressing South American tribes, degrading the environment, and engaging in ethnic and international conflicts. It would seem that sapiens is afflicted by an agressivity that exceeds its purpose. Where will it all lead?

Reflecting on the dehumanization process, Bernard Heuvelmans concluded that it must have taken place at all levels of development of the hominoid primates. He saw in it the "normal direction of evolution," expressing his deduction in a pithy statement that sounds like a "law": "To dehumanize, one had first to be somewhat human." It then follows that the diminutive common ancestor of both hominoids and anthropoid apes must have had human characteristics.

According to doctor of science and specialist of foot anatomy Yvette Deloison, that original ancestor would have been only some 20–30 inches ("a few tens of centimeters") tall:

---

2. Heuvelmans. *L'Homme de Neanderthal...*, pp. 459–460.
3. Heuvelmans, op. cit. p. 462.

Man has remained the animal closest to the original ancestor. He has refined a bipedal stance and equilibrium specific to his kind, and his foot has evolved in a unique fashion, with a strong arch and large big toes. Since his very origins, his hands have not been used for support, always remaining free and retaining their archaic structure.[4]

In other words, man is the descendant of an ancestor that was already bipedal.[5] The great apes diverged from that mode of motion; by using all four of their limbs, they "dehumanized." It is then the apes that descend from man (or at least from a man-like ancestor) and not vice versa!

The protohominid common ancestor would have lived in marshy areas, as do today's bonobos. Leading an amphibious rather than fully aquatic life, it would have taken advantage of the rich resources offered by the combination of wet and dry environments in a littoral zone, somewhat like today's sasquatches in the Pacific Northwest.

For zoologist François de Sarre, the sasquatch is similar to a tall variety of *Pithecanthropus*, close to the *Meganthropus* discovered in 1941 in Java by Dutch anthropologist Ralph von Koenigswald. As big as a gorilla, it is the fossil that most resembles today's bigfoot.

These amphibious *Pithecanthropes* left their littoral environment to take refuge in coastal forests; some ventured farther inland. François de Sarre interprets a hominid couple carved in a bone (less than 15,000 years old, found in the Isturitz Cave, French Basque Country) as a pair of swimming aquatic *Pithecanthropes*.[6]

---

4. Deloison, Y., *La préhistoire du piéton*, p. 188. The author acknowledges the influence of Heuvelmans and François de Sarre.
5. The bipedal stance if of great interest to anthropologists. A partial skeleton, 4 million years old, has recently been discovered in the Ethiopian Afar area (see *Sciences et Avenir*, no 69, Avril 2005). From the shape of its ankle, it was deduced that this hominid was exclusively bipedal and not partly arboreal, like Lucy (3.2 million years) or Orrorin (6 million years—the oldest known hominin related to modern humans, found in the Tugen Hills of Kenya in 2001; cf Wikipedia).
6. For François de Sarre's thesis on dehumanization and initial bipedalism, see http://cerbi.ldi5.com/article.php3?id_article=128.

The theory of original bipedalism is at first surprising. It goes against common preconceptions and, speaking personally, I long considered it with some misgivings. Nevertheless, Heuvelmans drew on solid scientific sources. The idea was advanced originally in 1924 by Berlin anatomy professor Max Westenhöfer who placed man's ancestor at the very beginning of mammalian evolution. In 1926, he characterized the most ancient mammal through its:

— pentadactily (having five fingers and five toes on each hand/foot);
— dentition in the form of a semicircular arc;
— adaptive nonspecialization (anatomical polyvalence);
— a bipedal gait.

Later, in 1936–37, the Russian–Belgian mammologist Serge Frechkop also advocated the concept of initial bipedalism. "He was convinced that the human foot had never passed through the evolutionary stage of a monkey's foot."[7]

Bernard Heuvelmans was a student of Frechkop and, pursuing his master's work, thought that quadrupedal and aquatic mammals had evolved from bipeds who had evolved away from bipedalism. "Modern Man has developed from the original bipedal ancestor," adds François de Sarre.

It follows that *Australopithecus,* among others, is not our ancestor; that Neanderthal is a specialized man, unrelated to monkeys, but descended from *Homo sapiens.*

As to the theory that says man was originally an aquatic creature, it is even more unsettling. That is certainly how I looked at it, in spite of the interest that I took in Bernard Heuvelmans' work. Then, I read the works of François de Sarre and other articles published in his magazine *Bipedia.* Finally, it was Yvette Deloison's book and her discussion of the Aquatic Ape Theory proposed by English professor A. Hardy that led me to change my mind. I can only invite the reader to delve into those authors' works—ready to bet that they will not be left indifferent.

---

7. See François de Sarre's website.

It also appears that a simple interpretation of Darwinian evolution as a linear process, often described as a parade of ever more human hominids, starting with a *Pliopithecus* monkey and culminating with *Homo sapiens* (sometimes carrying a briefcase), no longer makes sense. The parade as well as the family tree should be replaced by some other representation of evolution, but which?

Yvette Deloison suggests this simple diagram as a general framework.

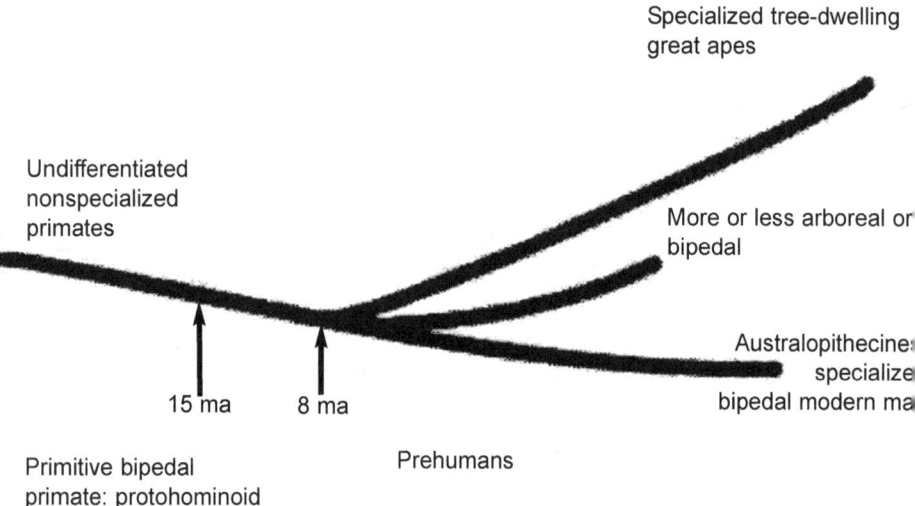

It was thus high time, following Bernard Heuvelmans and a few of his followers, to define new ways of looking at evolution, beyond random mutations or natural selection.

François de Sarre proposes a more detailed diagram, with extended explanation.[8]

The reader is asked to refer to the color chart on page 257 for which the following explanation applies.

---

8. De Sarre, *Homo floresiensis* : A little Woman on Flores Island gives Evolution its Right "Sense!" *Bipedia* n°24, <http://pagesperso-orange.fr/initial.bipedalism/24h.htm#8>.

EXPLANATION FOR COLOR DIAGRAM ON PAGE 257

Only the principal lines are represented here:

YELLOW: ancestral lineage *Homo sapiens*
BLUE: *Homo floresiensis*
RED: *Homo Neanderthalensis*
PURPLE: *Homo heidelbergensis*
ORANGE-PINK: *Homo ergaster–Sinanthropus–Homo erectus*
GREEN: Meganthropus

Throughout Pleistocene, the ancestral stock of *Homo sapiens* is characterized by the maintenance of the embryonic disposition of a round head, by an upright posture of the body, and by a frontal cortex above the ocular orbits (flat face and high forehead).

Since always, the species *Homo sapiens* admitted a broad intraspecific variability. Thus, the various dehumanized lines branched out; at the beginning, they are simple varieties of *Homo sapiens*. The distinct morphological features are explained by progressive modifications in ontogenetic process, during growth. Such modifications are fast, and correspond to a punctuation (as defined by Stephen J. Gould), probably after major cataclysmic events, that caused a separate evolution.

Homo erectus, Neanderthalensis, Floresiensis and others, are the results of an evolutionary process of hypermorphose, gradually spread in time. This reinterpretation of the evolution of *Homo* shares together with punctualism and gradualism.

New varieties, like *Homo floresiensis,* developed in geographic isolates, and resulted indirectly from pressures of the environment (exaptation). They are finally developing in a new species (speciation), then remaining in standstill (morphological stase).

Such robust forms can have survived until present time. They are the "wild men," whose vernacular names are put on the right of the diagram: ebu gogo, barmanu, almasty, sasquatch.

Our survey of the scientific literature pertaining to the wild man has revealed a rather forbidding situation. Even Marie-Jeanne Koffmann's analysis of the almasty's vision might be thought difficult to absorb. I wouldn't be surprised if the reader had forgotten that, in this part of the book, I have summarized a number of long and complicated works in order to draw out their essential conclusions. I have also thought it necessary to demonstrate the attention to details involved in scientific work. The persistence of scientists in their efforts to overcome confusion and arrive at an understanding consistent with the facts deserves that these facts and their conclusions be reported as they have presented them.

Grover Krantz's publications on bigfoot required consideration of the characteristics of the great apes, the hominids and Neanderthals in order to distinguish them from bigfoot, who would appear to be a relic giant primate, *Gigantopithecus blacki.*

Professor Porchnev's works would have quickly raised the eyebrows of Soviet leaders. How was the wild man to be shoehorned into the Darwinian credo favored by the communist regime? How to admit that the wild man was suited neither to work nor to discipline. The creatures described by Porchnev are suggestive of near-humans—large bipedal apes.

Such a conclusion was most surprising 30 years ago; it implied the survival of a Neanderthal, of a family of near-humans different from both *Homo sapiens* and the great apes. The recent discovery of Flores man has changed everything. Skepticism is no longer so fashionable.

The study of the wild man leads almost inevitably to efforts to find its place within the genealogy of man. We have seen, however, that there are as many genealogies as there are anthropologists! The latest and technically most advanced, based on genetics, is to be taken with caution (according to Marcel Otte); on the other hand, the rarity of fossils and the random nature of the fossilization process cast doubt on the reliability of classical family trees.

A few years ago, biologists could believe that Darwinian selection had replaced God as the ultimate reality, a dogma now being displaced by genetics and information theory. However the work of cryptozoologists, particularly Bernard Heuvelmans, has brought into focus two aspects of evolution. First, the existence of animals thought to be "impossible" within a traditional framework: the sur-

The Committee of the Praesidium of the Academy of Sciences of the USSR for the Study of the Question of the Snowman (1958). It remained officially in function for ten years. From left to right: Prof. P.P. Smoline, Prof. B.F. Porchnev, Prof. A.A. Machkovtsev, Dmitri Bayanov, and Dr. Marie-Jeanne Koffmann. According to Prof. Porchnev, the hairy wild men of Asia, including the yeti, were relic Neanderthals. This photo was taken in January 1968.

vival of the coelacanth, the incredible puzzle of the platypus. Second, the need to accept that Darwinism—or neo-Darwinism—tends to ignore rapid changes leading to speciation. Neither Darwinism nor genetics readily takes into account the phenomenon of hybridization, and the very idea of regression, or devolution, has been seen with great apprehension by scientific orthodoxy. However, the idea of initial bipedalism, as proposed by Frechkop and adopted by Bernard Heuvelmans, has become an inspiration to younger researchers (for example, François de Sarre and Yvette Deloison).

To declare, as did Heuvelmans, that the ape descends from man will also shock the creationists, sworn enemies of Darwinism. But imaginative young minds will soon know how to take advantage of a hypothesis which is paradoxical only in appearance.

Cryptozoology has facilitated a healthy reexamination of scientific thought. It seems to me that the interaction of the different disciplines that join within its fold provide a guarantee of serious objectivity. Every dogma is a blindfold. Cryptozoologists and more conventional scientists join forces in thinking and trying to prove that one has to go beyond the combination of random mutation and natural selection; that biochemistry alone can hardly explain certain "passages." They agree in thinking that there are many principles and forces at work within the vast skein of evolution.

No one could deny that it has been quite a struggle to come to understand the nature of the wild man! Even if there were any doubt about the scientific validity of the research undertaken in different countries, an important fact remains certain: the growing interest, manifest in all classes of society, in research on the wild man. Even as perhaps only a myth, it remains a powerful stimulus for scientific investigation.

For example, as a temporary conclusion to these chapters, consider this assignment given to the students of the University of Arizona. Their professors imagined that the creation of a complete skeleton of a nine-foot-tall (2.7 m) wild man would be a useful paleontological, anatomical and technical learning exercise. The result was a giant skeleton later to come into the hands of Christopher Murphy via David Hancock. This fabricated giant is now a stimulus to the imagination of researchers as well as for the general public.

Myth, the reflection of a nearly forgotten past, often hides enough revealing features to attract the interest of scientists. Their work, in turn, feeds everyone's dreams.

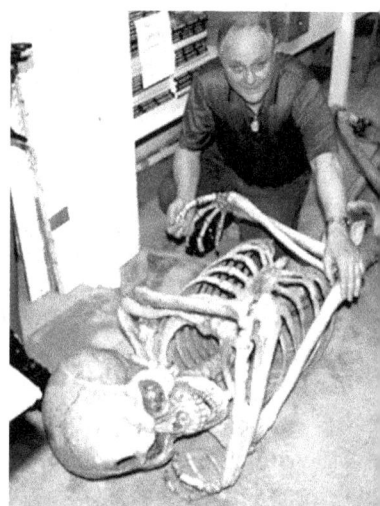

Chris Murphy with a fabricated "giant" skeleton.

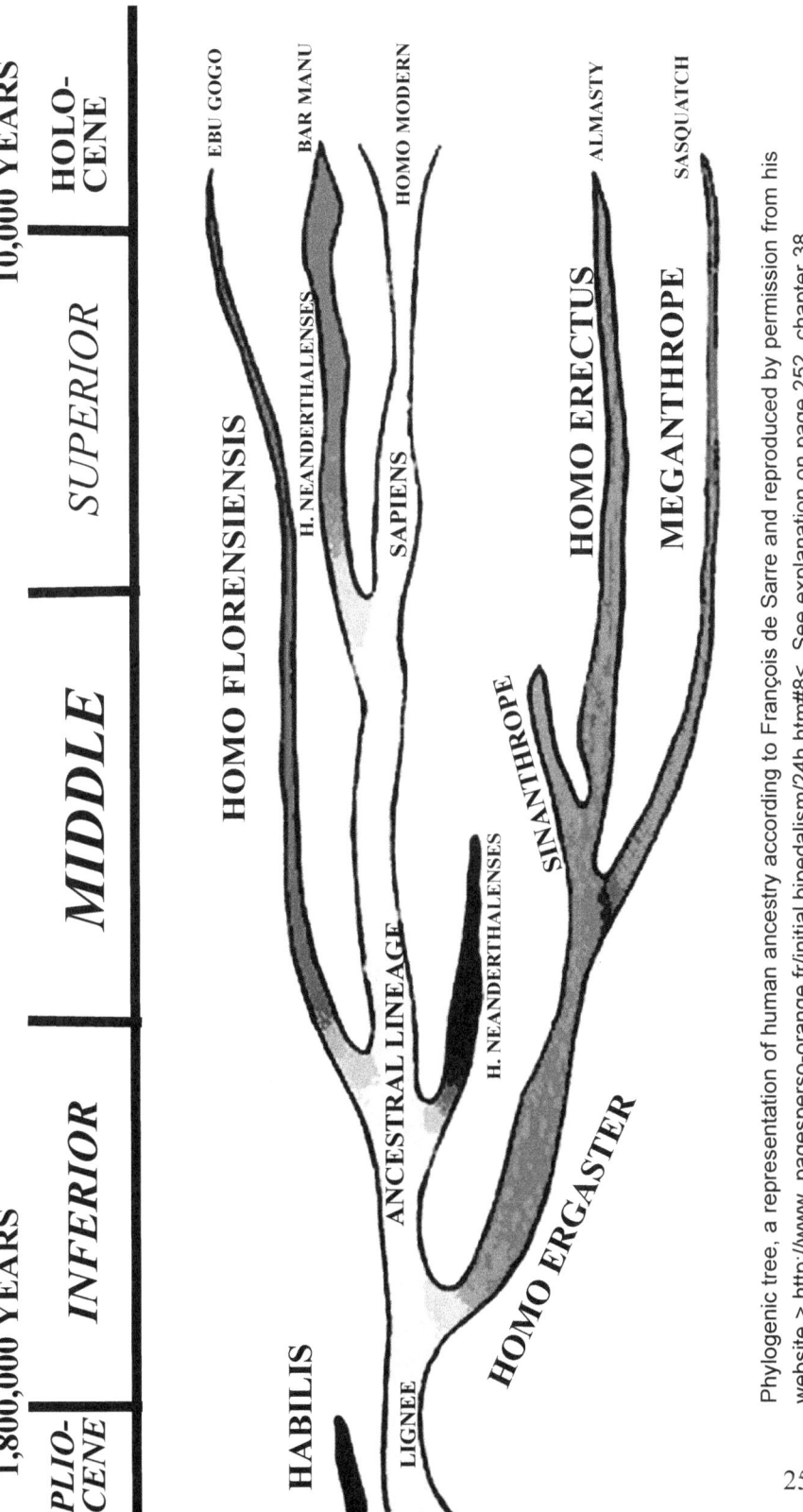

Phylogenic tree, a representation of human ancestry according to François de Sarre and reproduced by permission from his website > http://www.pagesperso-orange.fr/initial.bipedalism/24h.htm#8<. See explanation on page 252, chapter 38.

Peter Byrne (left) and the author at Bennett Pass, near Hood River, Oregon (July 22, 1995).

Frame 352 of the Patterson/Gimlin film. This film, taken October 20, 1967 at Bluff Creek, California, remains the main photographic evidence of the existence of sasquatch/bigfoot.

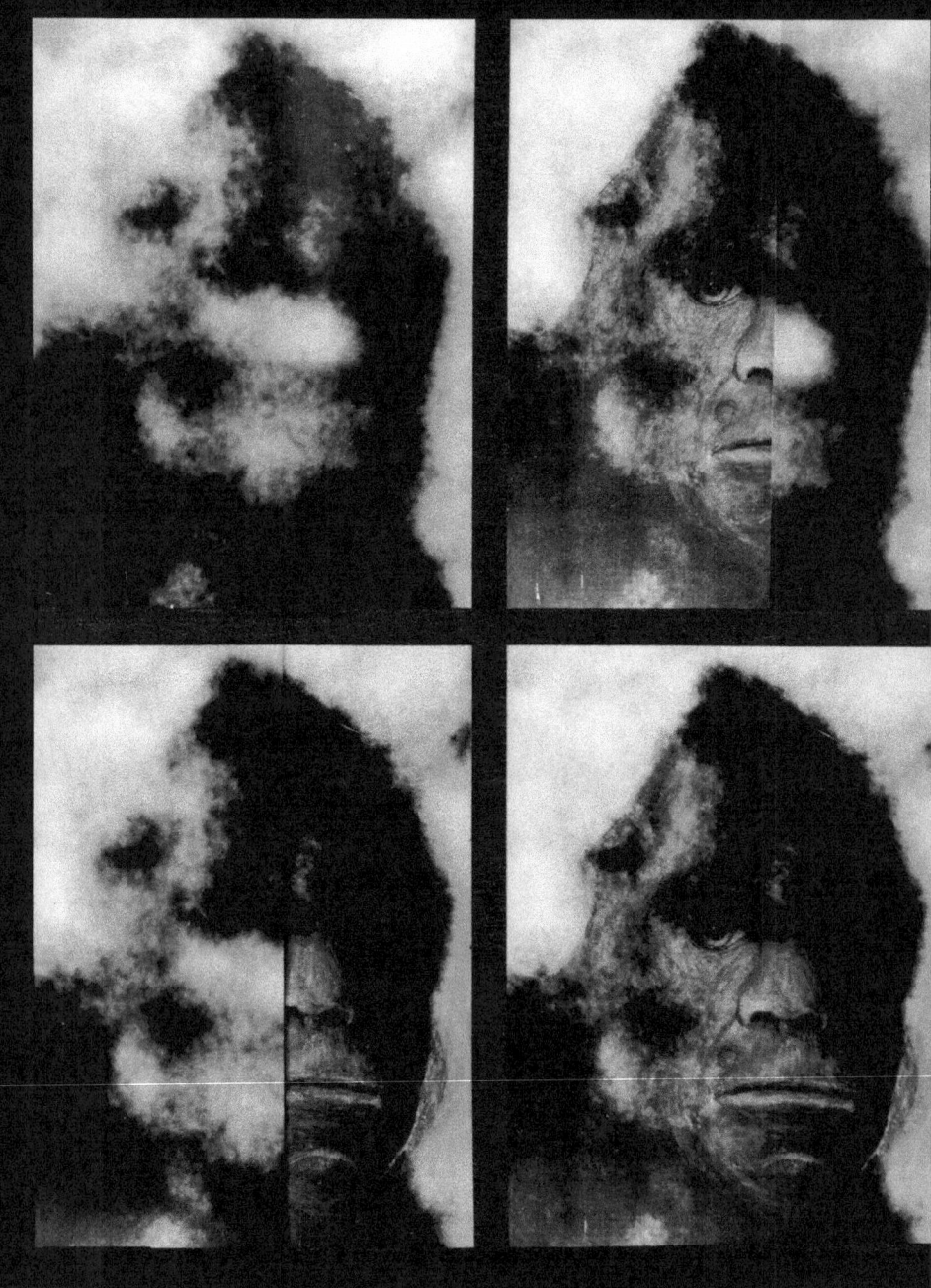

Artistic study of the Patterson/Gimlin film creature's facial features by Chris Murphy (1996). Top left is an enlargement of the head seen in frame 352. This image was enhanced using pastels to produce the subsequent images. The mouth was changed to a closed position to make it appear more natural.

(Top) Casts taken by Roger Patterson of the footprints left by the creature he filmed. The footprints measured about 14.5 inches (36.8 cm) long.

(Lower) The Bossburg "cripplefoot" casts, taken from prints found at Bossburg, Washington, in 1969. Note the deformed left (facing) print/cast. The left print was 16.75 inches (42.6 cm) long, and the right print 17.25 inches (44.5 cm).

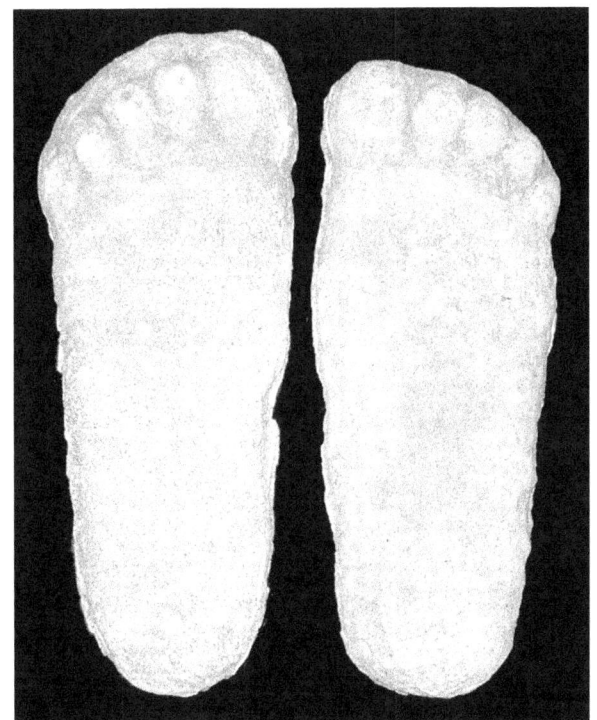

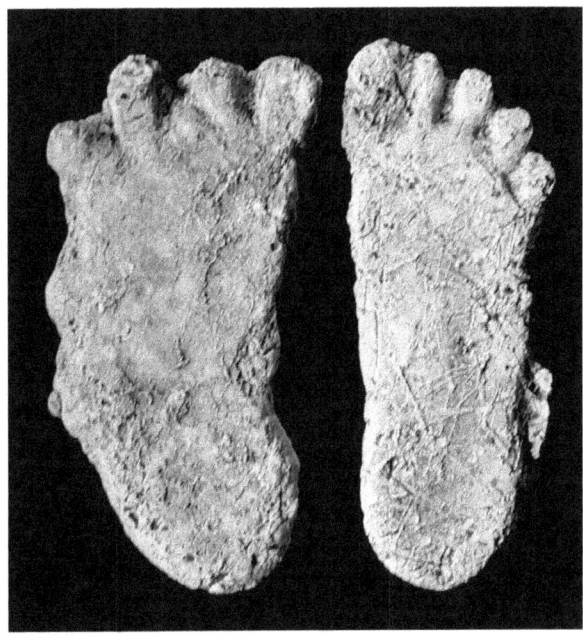

Bob Titmus is seen here displaying casts of footprints found along a Skeena River slough (near Terrace), British Columbia in July 1976. Young boys found the prints and Titmus investigated the find. The casts measure about 16 inches (40.6 cm) long.

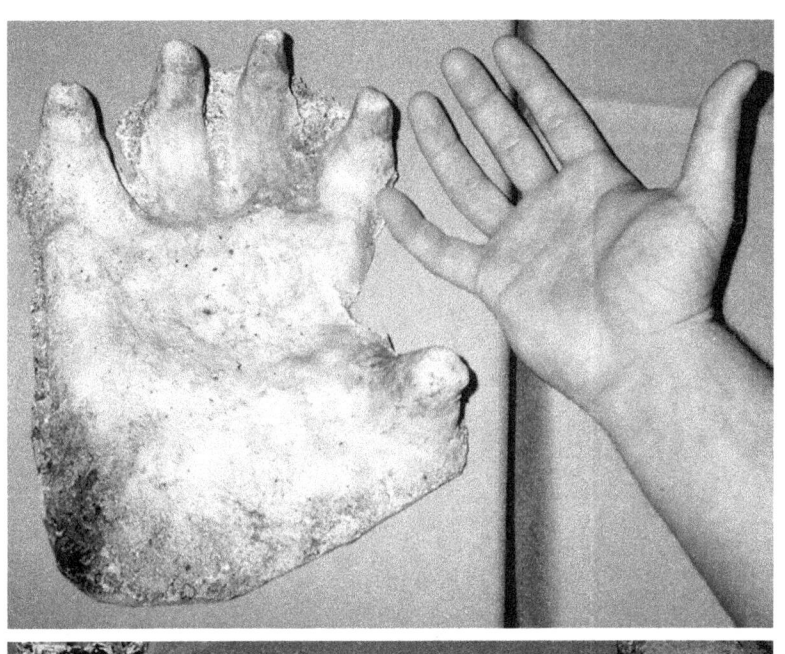
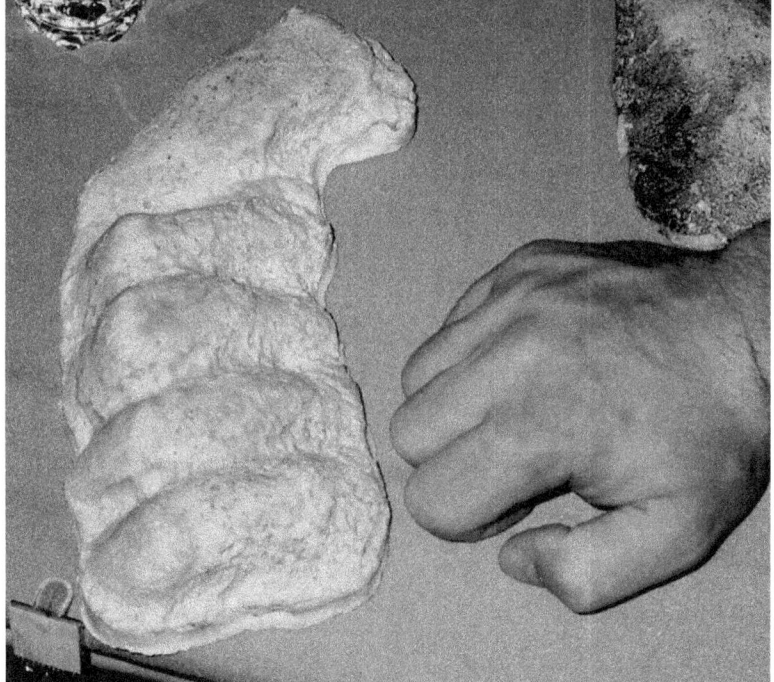

Casts of a hand print and knuckle print found in the Blue Mountains, Washington by Paul Freeman. The human hand shown is of a man about 6 feet (1.83 m) tall and about 200 pounds (90.6 kg).

The intriguing Skookum Cast, with Dr. Jeff Meldrum. The cast was made from body imprints believed to have been left by a non-human primate of some sort (assumed to be a sasquatch/bigfoot). It is believed the creature reclined in soft earth and reached over for fruit that had been placed by researchers in an attempt to get footprints.

William Munns with his famous scale model of a *Gigantopithecus blacki*. The model is currently on display at the University of Iowa.

The tamed Columbia River under summer skies near the Grand Coulee Dam.

Bronze statue of Gerald Durrell (1925–1995) and lemur friend, Jersey Zoo, Jersey, Channel Islands. G. Durrell was a naturalist, author and pioneer animal conservationist.

Zana, the Russian Almasty woman with one of her babies. This astounding image was created by Brenden Bannon, a professional artist. Zana had several children by ordinary men and descendants are still alive.

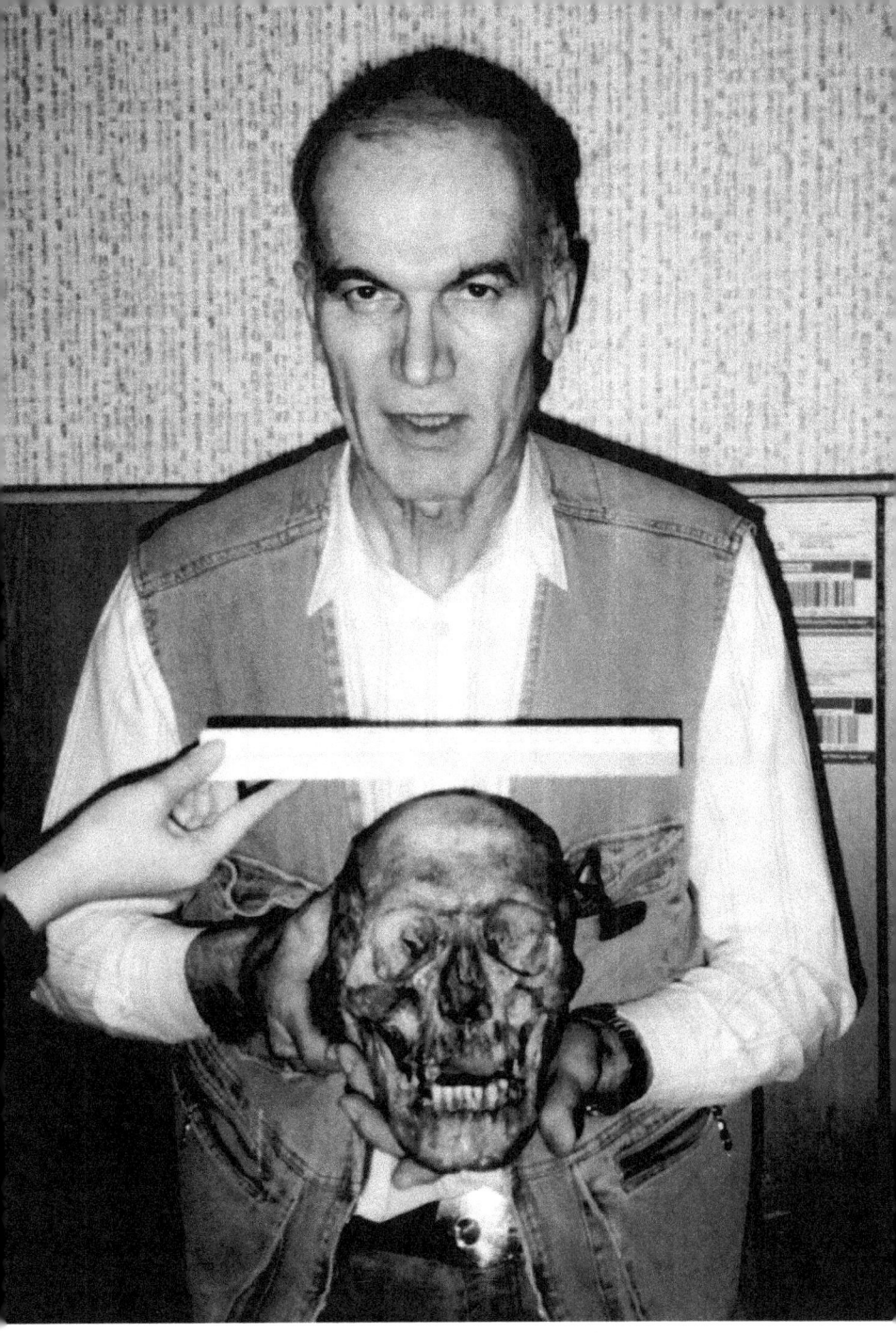

Igor Bourtsev is seen holding the skull of Khwit, one of Zana's sons. Analysis of the skull is inconclusive as to the nature of Zana—scientists are unable to agree; however, DNA results indicate that Khwit was human.

Julien, the author's son, with pandas at the Wolong Giant Panda Reserve Center, Sichuan, China (April 2006). Young pandas are playful, but adults eat and sleep most of the time.

Dzonokwa: Kwakiutl sculpture, Museum of Anthropology, University of British Colombia, Vancouver, Canada. The prominent fang is a sign of the ferocious nature of ugly Dzonokwa.

Dzonokwa, a sculpture by Bill Holm (circa 1990), Burke Museum, Seattle, Washington, USA. Bill Holm is a renowned sculptor and academic commentator on Pacific Northwest Native art.

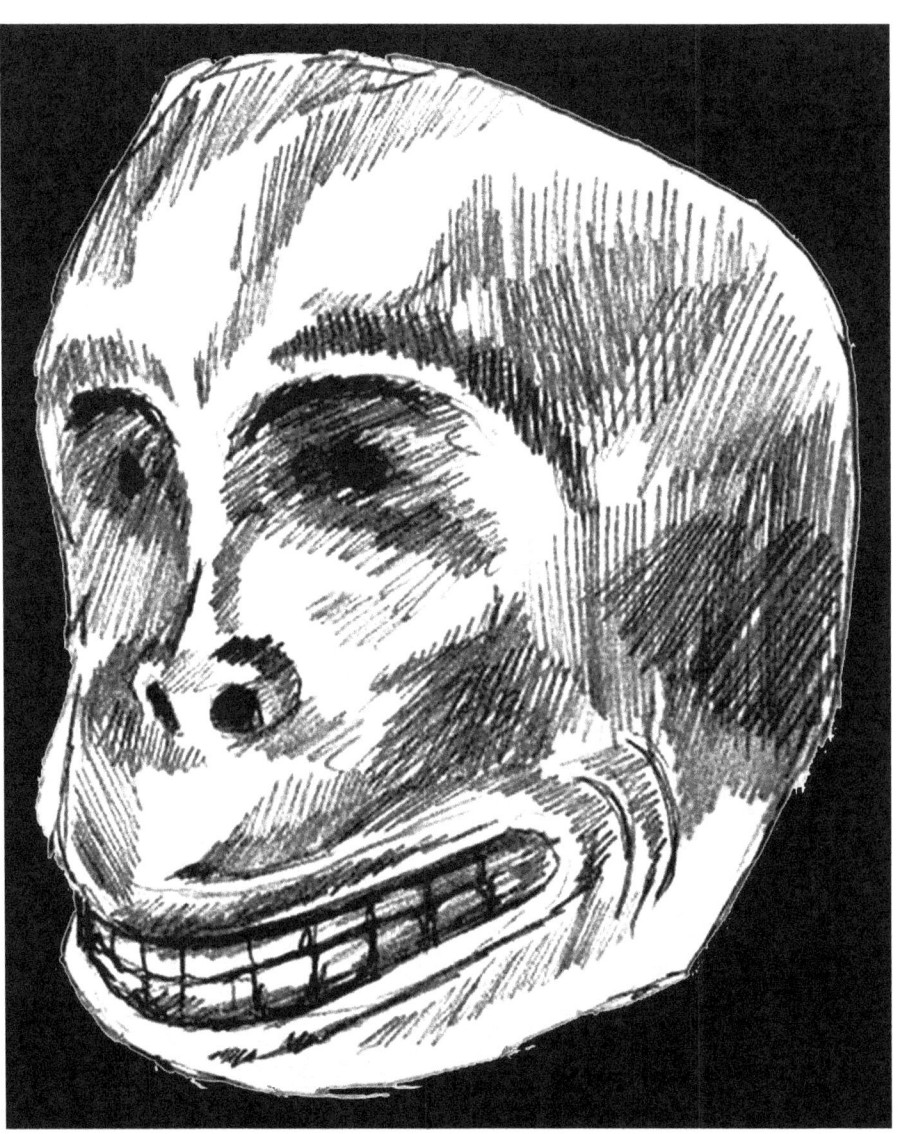

By far the most revealing of all Native sasquatch masks is this Tsimshian mask discovered in British Columbia in 1914. It likely dates back to the mid 1800s. What it depicts is undeniably an ape-like creature of some sort. It is considered a very sacred artifact by the Tsimshian people. This drawing by Peter Travers was created from a photograph.

A Kwakiutl buck'was or "wild man of the woods," (copy of an old original mask). The buck'was is the spiritual embodiment of all that is found in the forest. He is rarely seen.

Gagiid, a mask by Ralph Bennett (1995), author's collection. The Gagiid dance starts with a wild component; in a second part, Gagiid is captured and recovers his human nature.

(Above) Raven (1994). (Below) Killer Whale (1994).
Two pieces by Haida carver Ralph Bennett whose Native name, Goolaslacoon, means "Abalone Fingers."

(Above) Ralph Bennett (1995).
(Below) The carver's hands.
The city of Nantes, where the author taught at the university, has two totems by Ralph Bennett.

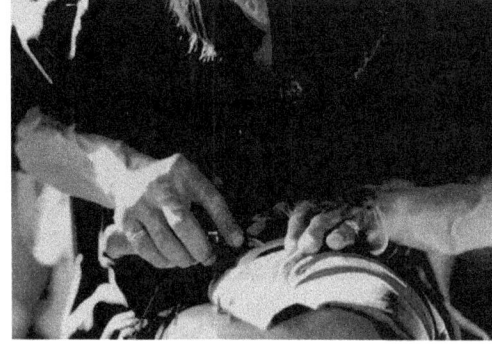

An elaborate Kwakiutl Dzonokwa (sasquatch) transformation dance mask. The mask opens to reveal different figures (heads).

Native dancer Tsungani with the transformation mask (see previous page) and associated costume. (Chief Don Assu Potlatch in Campbell River B.C., 2002.)

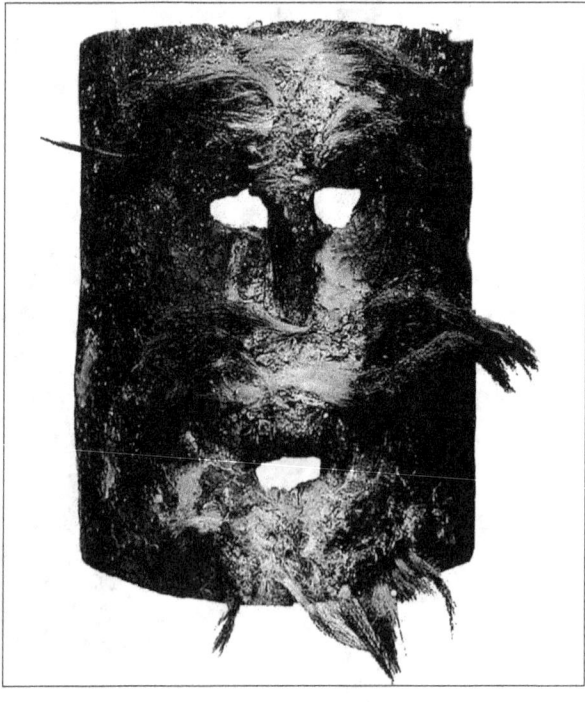

(Above) Fred Bradshaw (1947–2004) shows his sasquatch mask (Quinault) and the wooden ball used to delineate the "tribe's circle." He is also holding a sprig of wild sagebrush (the equivalent of the incense burnt in Christian churches). (Below) Wild man mask (Nepal).

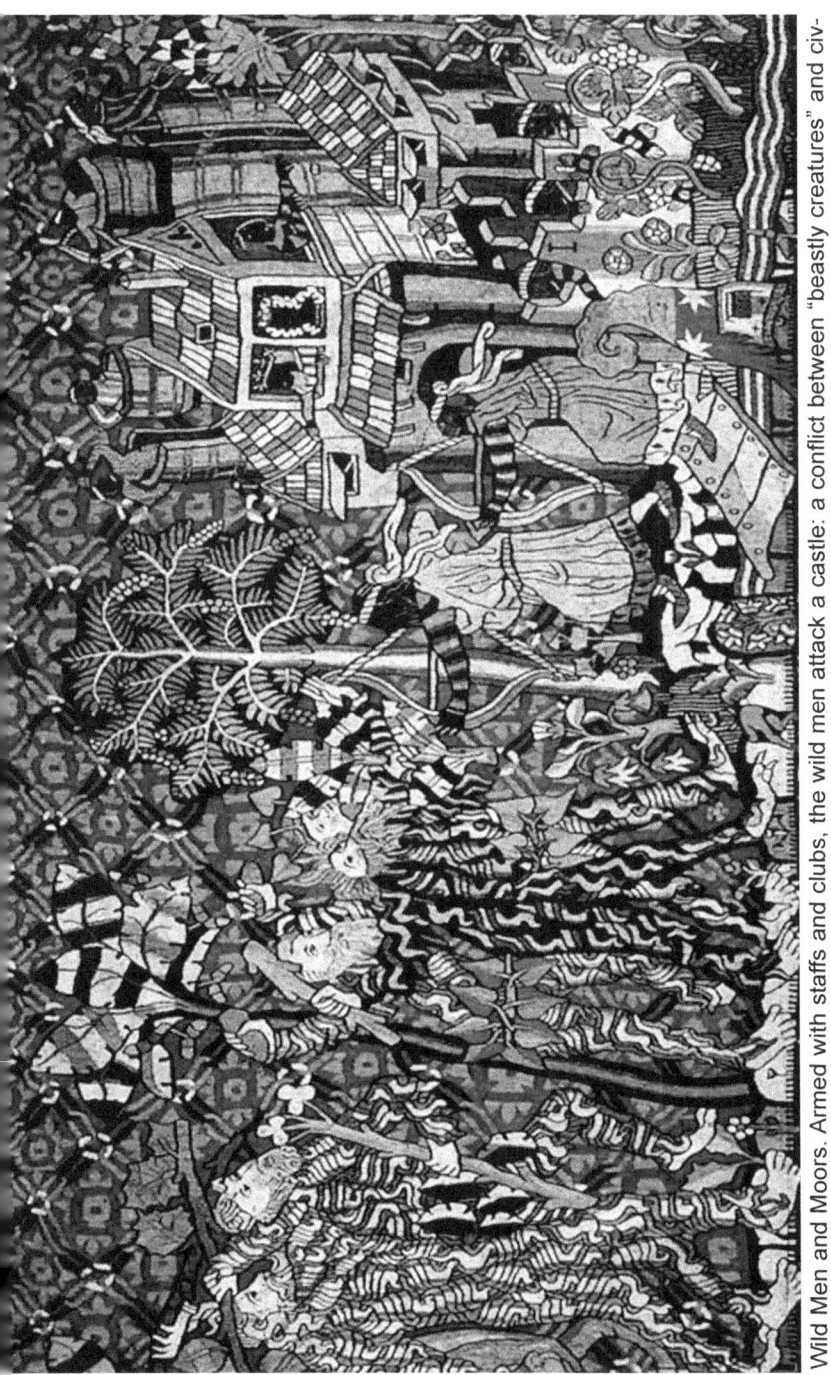

Wild Men and Moors. Armed with staffs and clubs, the wild men attack a castle: a conflict between "beastly creatures" and civilized people—a common theme during the Middle Ages and Renaissance. The tapestry author is unknown (circa 1400, Museum of Fine Arts, Boston, Massachusetts, USA).

Raven and the First Men (1983), a masterpiece by Bill Reid (1920–1998), Museum of Anthropology, University of British Columbia, Vancouver, Canada. The sculpture illustrates the Haida creation story, but also "Raven is perched on the clam...Above him one should imagine another being...He is called Sasquatch" (see chapter 51).

Wild Man: a fresco on a Buddhist Temple of the Daizu (Dai ethnic minority), village of Menghai, Yunnan, China.

The wild man of China, or yeren, has been sighted in the Shennongjia Region, Province of Hubei, more often than anywhere else. These mountains with their dense forests provide shelter for the mountain goat, golden monkey and leopard.

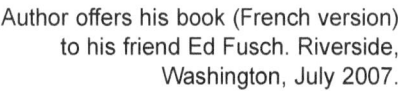

Ed Fusch, goldminer, prospection equipment dealer and "philosopher extraordinaire," demonstrating a mercury retort in his Riverside, Washington shop in 1995. The mercury itself is used to produce an amalgam, then a retort is needed to separate the mercury from the gold.

Author offers his book (French version) to his friend Ed Fusch. Riverside, Washington, July 2007.

Ray Crowe, editor of *The Track Record*, photographed by author in the "basement museum" of his bookstore in Hillsboro, Oregon, 1994.

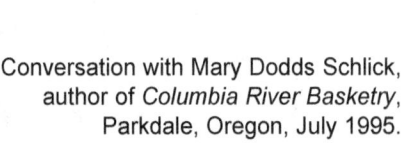

Author (left) with researchers Fred Bradshaw and Wayne Moore in the latter's computer repair shop, Montesano, Washington, July 1994.

Conversation with Mary Dodds Schlick, author of *Columbia River Basketry*, Parkdale, Oregon, July 1995.

The author (left) with Paul LeBlond at Paul's home, Galiano Island, British Columbia, August 2007.

Discussion on the wild man with a Sichuan archeology professor, Chengdu, China, February 2008.

Lecture on the wild man with interpreter Ao Min, Sichuan University, Chengdu, China, February 2008

Bernard Heuvelmans, the Father of Cryptozoology, guest of honor at the annual convention of the CERLI (Centre d'Etude des Littératures de l'Imaginaire), University of Nanterre, France, March 1992.

PART IV

Sasquatch and the World of Mythology

"It would be wonderful to be able to accompany Ed Fuchs interviewing the Indians about Bigfoot: they are the only ones I trust in this business."

BERNARD HEUVELMANS
(Personal Communication, April 1992.)

*"A Cannibal has come down from the mountain, Raven,
And preys upon our village.
Only a few of us are left."*

NICK DIMARTINO
(Seattle playwright, *Raven,* 1975.)

Chapter 39

A Bridge Toward Mythology

The following pages are concerned essentially with myth. It couldn't possibly be otherwise since pre-Columbian America was inhabited by tribes who lived within a flora, a fauna and natural surroundings that were endowed with spirits or, if one prefers, souls. Their spirit was reflected through costumes, everyday utensils, weapons, masks, sculptures and stories.

The best way to understand even a fragment of these captivating myths is to remain silent. That is indeed the nature of myths: their external form is only a minute part of their essence. Their appearance is highly variable and they demand long-term patience on the part of the researcher.

Such an attitude of restraint and humility is eventually rewarded, especially if the researcher surrenders to the fascinating sceneries and fantastic roles played by the spirit-actors. The bear, the whale, the eagle and the coyote playing their part within their natural surroundings also stand for the spirits that they represent and which transcend their physical presence.

Let us admire the totems, the basket decorations, the dance; let's even join the dance if we are invited. We shall gain a deeper understanding.

Cryptozoology offers exactly the right transition between science and mythology. Let's turn back for a moment to Bernard Heuvelmans' works, particularly those relating to the ancient stories of wild men.

By the way, the creation of the International Society of Cryptozoology in 1982 in the USA, presided by Bernard Heuvelmans, with professor Roy Mackal as vice-president, led by example to the creation of a number of other associations. Some of these might have had modest goals, but often suffered from a lack of intellectual ambition. We have seen earlier how the ISC associated to its ranks a number of famous honorary members. The quality and variety of the contributions to the ISC Newsletter and to its annual refereed journal, *Cryptozoology,* the multidisciplinary nature of its symposia, often organized in the UK, naturally spawned many envious imitators.

Bernard Heuvelmans wrote a number of fundamental contributions in the pages of *Cryptozoology*. As did many other readers, I found his work inspiring, stimulating the wheels of my imagination. Let's consider for example Heuvelmans' list of unknown land creatures:

> In Europe (Palearctic bioregion): Hairy wild men, probably Neanderthals having survived as late as the historical period. Known in antiquity as satyrs—a word borrowed from the hebrew se'ir (hairy)—and known in the Middle Ages as wudewása (woodland creature), it is reported as late as the 13th century in Ireland, up to the 16th century in Saxony and in Norway, in the 18th century in the Swedish island of Öland and in Estonia, and until at least 1774 in the Pyrenees (as iretjes, or basajaun), and finally in the Karpathian mountains (the wild man of Kronstadt— today Brasov, in Rumania) until 1784.[1]

This list refers to the wild and hairy men of the Old World. The satyrs take us back to the earliest satyr of them all, the god Pan, first seen in imagery of the fifth century BC. Bearded and hairy, mainly below the belt, he had forked hooves, a flat nose and little horns on his forehead. He is first shown naked, but later appears dressed in the skin of a beast, usually a lynx. He holds a syrinx in one hand and a stick in the other. Syrinx was a nymph, a companion of Artemis (the Diana of the Romans), who rejected Pan's assault on her virginity. The nymphs of the river transformed her into a field of reeds. Pan made his flutes from these reeds and called them syrinx (today's Pan flutes). The carnal appetites of the god of the pastures are legendary. It is he who oversees the fertility of the herds. He is said to be the source of those noises and cries emanating from the valleys and forests which caused panic in people and their flocks. Legend also has it that at the time of the Crucifixion, when the temple's veil

---

1. Heuvelmans, B. 1986. "Annotated checklist of Apparently Unknown Animals with which Cryptozoology is concerned," *Cryptozoology,* 5, p. 12.

was said to have been torn, a great cry carried all the way to the ocean to the ears of a sailor: "The Great Pan is dead!"

Pan is the only god reported to have died.

Pan had a son, Silenus, also pug-nosed, and sporting the tail and ears of a horse. He was Dionysios's tutor (Bacchus in Rome). Silenus in turn had sons, the Silenoi, famous for their pranks, similar to their father without having inherited his wisdom.

Among Romans, Faunus was the equivalent of the Greek Pan. Like Pan, he came from Arcadia, a province of the Peloponnese which stood for the peace and happiness which accompany a pastoral existence. The satirist Juvenal (first century AD) made fun of the Arcadians, using the term *arcadicus juvenis* (Arcadian youth) as a synonym for stupid country bumpkin. Nevertheless, Faunus' powers of prophecy were in high demand. Rabelais also thought of Arcadians as simple-minded.

Rabelais created the giant Pantagruel (from the Greek *panta* = all, and Arabic *gruel* = thirsty), ever searching for the oracle of the Divine Bottle. For most of us, Pantagruel was the last of the Old

Nebuchadnezzar—drawing by William Blake (1795). The Chaldean king dreamed that he cut down the unifying axis, the World Tree. Henceforth, his human heart was replaced by that of a beast.

World's giants, while the New World remained haunted by gigantic creatures, ogres and cannibals. A New World where a particular tall creature, the sasquatch, excites scientific curiosity and figures as a mute oracle of a spiritual renewal. Is Pan being reborn?

Bernard Heuvelmans reminded us that the wild man was a recurring figure in the Middle Ages. Indeed, he appears everywhere in the art and literature of the times: in Arthurian novels, in the epic poems of the German minstrels, in the 16th century works of Cervantes and of Spencer.

The wild man figures in the paintings of Pieter Brueghel and Albert Dürer; he shows up on tapestries and on the jewel cases offered by gallants to their ladies; he decorates saddles and suits of armor; he appears on the ceramic tiles of stoves, on candle holders and tankards, and on the carved end of roof beams. He is present on church buildings, on the capitals of columns, on baptismal fonts, tombstones, and in gargoyles. In one Spanish church, he takes a place normally reserved for saints, on the risers of the main portal.

The wild man is featured in well known writings. For example, in the twelfth century novel *Yvain ou le chevalier au lion*, by Chrétien de Troyes, where he is described as a human being, although repulsive in appearance. Similarly, Heinrich von Hesler explains in his fourteenth century Apocalypse that the wild men are children of Adam by their form, visage and human intelligence, and that they are the work of God, although they live far from human habitations.

French history books tell the story of the Bal des Ardents, which took place in 1393 at the court of king Charles VI. At that time, "hirsutes"—wild men perhaps—were exhibited in popular fairs; they were later replaced by bears. Charles VI and his courtiers had disguised themselves as wild men, outfitted with costumes made of linen cloth sewn onto their bodies and soaked in resinous

Merlin the Enchanter — bronze wall vase (ca. AD 600). Merlin is shown transforming into a salmon. Among the Welsh, Merlin became the archetype of the magic fish of Celtic tradition.

wax or pitch to hold a covering of frazzled hemp, so that they appeared shaggy and hairy from head to foot. One of the costumes caught fire and spread to the other dancers. The king was saved by the presence of mind of a lady who wrapped him in her robe, thus stifling the flames.

The sixteenth century English poet Edmund Spenser, in his *Faerie Queene*, mentioned that these creatures lived far in the forest, near a remote clearing avoided by wild animals.

All these creatures from the pre-Christian world, known as "savages" (from "silva," the forest), are intimate with the secrets of nature. Occasionally, they reveal snippets of their knowledge, advising peasants about the weather to come, the yield of their crops, the use of medicinal herbs, and even how to harvest rye.

One of the wild men immortalized in literature is Merlinus Sylvestris, Merlin the Wild of Arthurian legends. Speaking of Arthur, we recall that *arctos* in Greek means "bear" and in Latin *arctus sumnus* (a bear's sleep) is a deep sleep. Merlin could transform into a child, or into an old man, even an animal, usually a deer. Dark of skin and hirsute, he knew the speech of animals. His behavior was often contradictory. As the old saying goes: "Like the wild man who sings when the weather is bad, I have decided never to stop singing." We know that the bear, after having left his den at the end of winter, can return to it and fall back to sleep. This behavior is only apparently contradictory: the bear foresaw a return to cold winter conditions. Observant peasants benefit from that kind of warning.

Similarly, Merlin is said to have laughed, apparently without reason, while watching a poor pilgrim repair his shoes before starting on a long trip. After a few miles, the pilgrim collapsed and died. To die so quickly with brand new soles on his shoes. What a joke! Merlin had foreseen the fate of the unfortunate pilgrim.

Merlin, son of the Devil, knew the past; son of a virgin, he also foresaw the future. He already spoke at birth and seemed to belong to another period, that of the Saturnalia, when humans temporarily recovered their ability to speak to animals.

In his twelfth century *Vita Merlini (The Life of Merlin),* Geoffrey of Monmouth describes Merlin as a wizard or a shaman, the true heir of Eurasian shamanistic traditions.

Just recently, in the small German town of Hombourg, near the

French border, François de Sarre discovered square pottery tiles in a castle dating from 1550. On these tiles are shown images of two types of wild man. According to de Sarre's description, one of them shows a very hairy *Homo sapiens*, while the other shows a creature with an upturned nose, also with a thick head of hair, similar to Heuvelmans' pongoid man. Feral modern men would thus have coexisted with survivors from the Paleolithic era.

In his book *Antiquities of the Jews,* first century Jewish historian Flavius Josephus offers a description of John the Baptist as a very peculiar man who had glued animal hair to those parts of his body that were not naturally hairy so that he looked like a wild man.

Fifteen centuries later, the German painter Schweiger represented Mary Magdelen as a wild woman, her body completely covered with hair, except for her breasts, knees and elbows.

Often, the wild man (or woman) is seen merely as an animal resembling a human being. Because of his intellectual, moral and spiritual deficiencies he is thought to be a lunatic or a henchman of the Devil. Over the years, the Catholic church strove to expurgate all pagan symbols and teachings, as seen for example in this poem where seventeenth century English poet Robert Herrick, preaches the abandonment of idolatrous worship of the forest and its spirits.

Ceremony Upon Candelmas Eve

Down with rosemary, and so
Down with the baies and misletoe;
Down with the holly, ivie, all
Wherewith ye drest the Christmas hall
That so the superstitions find
No one least branch there left behind:
For look, how many leaves there be
Neglected there, maids, trust to me
So many goblins you shall see.[2]

Peasants' stories, transmitted by word of mouth, relate the comings and goings of allegedly cannibalistic *louberous* (werewolves)

---

2. *Hesperides: or the works both humane and divine of Robert Herrick, Esq.* Vol. II, p. 148, Little Brown & Co., Boston, 1856.

In 1890, Basque shepherds claimed that a Basa Jaun regularly came to beg for food and that it would occasionally steal some from their pens.

in the Alps, near Saint-Maurice en Valgaudemar. In Saint Maximin, it was said that the wild men who lived in the area were incapable of speech. When they came to the village, they would be given something, bread, for example, to get rid of them.[3]

Were these isolated cases, small marginal communities of woodcutters or charcoalers, or were they "living fossils" of the Neanderthal type? Genuine wild men had perhaps survived to the time when Louis XIV was having Versailles built.

One of the consequences of the abasement of wild men to the status of beasts was the depreciation of the value given to their life. Hunters took pride in killing what they felt was a blemish on God's creation. Nevertheless, it made them uneasy and they avoided boasting about it, preferring to claim to have killed a bear.

French postage stamp (issued in 1991).

Indeed, I have more to say about the bear, as it is closely linked to the wild man, as exemplified by Merlin. Even today, stories, feasts and carnivals carry on the ideas of the transformations between bears and wild men. They do the Bear Dance in Romania. Until recently, they danced the Bear Jig at family reunions in Canada. Such dances followed a common pattern, mimicking the hunt, the killing of the bear and finally its resurrection. After its death, the beast was brought back to life by blowing air into its anus.[4] The hunter would then skin the animal and shave it carefully. Back to life again—twice reborn— the bear would verify the quality of the shave by using a spectator's buttocks as a mirror, to universal laughter.

---

3. From Joisten and Abry, *Etres fantastiques des Alpes*.
4. When the bear wakes up from hibernation, it pushes out the anal plug which closes the intestine during the winter. Now ready to function, the animal's vital activities resume.

Forest men of Transylvania (Romania).

Every February, the people of Prats-de-Mollo (in the French department of Pyrénées Orientales) celebrate the Festival of the Bear *(El dis de l'os)*. The bears are played by young men who swarm all over the village, smeared with a sooty compound and dressed in sheep skins. Hunters chase them, shooting blanks. For a few hours, the bears have the run of the village, smudging young women, their favorite target, as well as the odd spectator. Finally, they are caught and shaved by white-coated barbers using an axe for a razor, blood sausage for a brush and wine for water. In that area, there were still bears until some time between the two world wars.

It would be tedious to list all the traditional holidays that illustrate the springtime transformation of nature, all echoing a similar rebirth in human beings. In Europe, such celebrations occur everywhere at the approach of spring. Our last example is the highly popular Urnach festival held in Switzerland, where the wild man appears covered with moss, bark and fir boughs. There, the wild man is closely linked with the Green Man, and at the same time, with the bear in a celebration of the return of the light.

Following Bernard Heuvelmans methodology, rumors, tales, and recent legends, as well as ancient mythological stories, are to be taken into account. This is exactly the method followed by Dr. Vladimir Markotic in his study of the God Pan.[5] Markotic's metic-

---

5. Vladimir Markotic, "The Great God Pan — an early hominid?" in *The Sasquatch and Other Unknown Hominoids*, pp. 251–264. Of course, the author ends his presentation with a question: "Was Pan one of the first hominids?"

Bigfoot portrait by the animal illustrator Stefano Maugeri (Rome, 1999).

ulous description gives full satisfaction to the student of mythology. The myths of interest to cryptozoologists are being critically and methodically examined. Such a reexamination of the great myths constitutes a valuable body of work for anyone trying to understand today's bestiary.

For Heuvelmans, the names given to these various beings differ of course between countries, but they usually have similar meanings, such as men of the forest or savages, etymologically equivalent appellations, satyrs, hairy or bear men, which is also the same thing. It would be a serious mistake to believe that these wild and hairy men were creatures of the imagination and that they all belonged to a single kind of flesh-and-bone being.[6]

---

6. Bernard Heuvelmans, personal communication.

Let us keep those historical examples in mind. They belong to the heritage of Old Europe, but we perceive equivalencies between European wild men and those that haunt the forests of North America. It is now time to return to the New World. Over the past 25 years, bigfoot investigators have prospered in North America. Observations have been gathered from Florida to the Pacific Northwest, as well as Ohio and Vermont.[7] It is of course on the Pacific Coast that most sasquatch sightings take place.

Our survey of sasquatch lore would not be complete without mentioning some rather far-fetched ideas. Most bigfoot hunters find the notion that the creature might have been dropped on our planet by extraterrestrials rather outlandish. It is advanced by a few individuals working from some eccentric premises. As a kind of poetic approach, it does exert a strange attraction.

A literary application of such theories sometimes generates enjoyable works, such as *Sky Man on the Totem Pole*,[8] a novel where Canadian writer Christie Harris blends folklore and science fiction. In this story, a legend of the Pacific Northwest tells of a strangely attired man abducting an Indian princess and bringing her back to Earth many years later with her six grown children. They return with a mysterious box capable of destroying everything. These "men of heaven" present themselves as gods to the Indians. However, the latter, with the help of their shaman acting as an intermediary between men and spirits, become a religious inspiration for the extraterrestrials. Thus, Indians and aliens enrich each other at the spiritual level.

This well-meaning novel is perfectly suited to adolescent readers. One might draw a parallel with some concerns expressed by Stephen Jay Gould. Given that modern biotechnology opens up the possibility of blending species, bringing together via gene insertions small parts of lineages separated by millions of years, how will one know when to restrain the application of a technology which offers immediate rewards? All organisms are emptied of their vital essence and transformed into abstract messages, a fact that I deplore, as does

---

7. The Wild Man seems omnipresent! For Australia, see *Bunyips and Bigfoots*, Malcolm Smith, Millennium Books, Alexandria, Australia, 1996.
8. *Sky Man on the Totem Pole*, Christie Harris, illustrated by Douglas Tait, McLelland & Stewart, Toronto, 1975.

Gould. That is precisely the illness that affects Harris's "space men." It is also what the Indians risk forgetting, she adds, advocating a return to a holier respect for nature.

Sculpture: House of Queen Anne, Duchess of Brittany, Morlaix, France, fifteenth century. The branch or tree trunk indicates a close link with nature, hence the name Woodwos (England), *Uomo selvagio* (Italy) or *Homme Sylvestre* (France), all meaning "man of the woods."

Chapter 40

The Literal Truth

In the minds of some people, there are legendary writings that are interpreted narrowly and literally, their words read as infallible and holding inerrant truth. This approach, as discussed by Stephen Jay Gould, leads "theological positivists" to finance expeditions, such as that to discover Noah's ark. In the United States, fundamentalists have created the concept of "creation science." According to their interpretation, the Earth is only 10,000 years old and all species were individually created by God. Creationists also oppose Darwin's theories, as, for example, in the 1925 Scopes "monkey trial." At that time, 20 states declared that to teach "man descended from an inferior animal" was a crime. Concerned with the modern resurgence of such fundamentalist views, Gould wrote, *Rocks of Ages: Science and Religion in the Fullness of Life*.[1]

Sasquatch investigators distance themselves from such arguments, which remain limited to the United States. Those of Christian faith make a clear distinction between their quest for the wild man and their religion. For others, probably the majority, creationism is irrelevant and of no concern at all. For some, both men and women, the sasquatch becomes a real obsession, invading their consciousness and commanding a large part of their time and energy.

As it is sometimes perceived, sasquatch reminds me of a remark made by George Schaller about there being two giant pandas, one of them existing in our mind, the other in its natural habitat.

The real panda is really boring. It spends 60 percent of its time chewing bamboo and the remaining 40 percent resting and evacuating an enormous mass of nondigested shoots. According to those rare sasquatch witnesses who have managed to observe the creature from a safe distance without alerting it, the real sasquatch carries out a much wider range of activities.

In the imagination of many inhabitants of the Pacific Northwest,

---

1. Translated into French as *Et Dieu dit: Que Darwin soit* (And God Said: Let there be Darwin), a rather more impactful title!

the sasquatch is a benevolent giant, deserving respect as it reigns over the forest. People benefit from its wisdom.

Those who would meet the sasquatch do not hesitate to venture in the woods, walking for hours or lying patiently in wait for its passage. They fear neither boredom nor failure. sasquatch field investigators also think of the sasquatch as the ambassador of wilderness, of unspoiled nature in its original vigor and beauty. How could one even think of failure? Even if a sasquatch does not show up, the seeker will have shared its kingdom for a few hours, or days, feeling also like masters of the forest.

Cover of George Schaller's book.

The quest for the sasquatch requires knowledge of its habitat, its place in the ecosystem, its habits, and its role in Native culture. Although the Native people I interviewed were fully aware of the object of my research, they persisted in refusing to provide the answers or even the clues that would have moved my search forward. However, they kept telling me about the practices formerly or currently followed by the Indians. They told me about the importance of traditional celebrations. During powwows, the participants, even the children, wear splendid costumes. Pride sparkles in the dancers' eyes; I have witnessed it myself.

Some dances, aiming at maintaining a brotherly link between all creatures, imitate the behavior of animals. The costumes are decorated with feathers, shells, furs, bones or claws, a reminder that animals as well as humans are part of nature. This teaching starts early in life. As a Blackfoot woman expressed it: "A very high type of civilized people, which is what we American Indians were. We are learning all over again to become civilized."[2]

The author adds that as far as "medicine" goes, it is important to

---

2. Nell Kipp, *Pre-Columbian History of the Red Man*, p. 39.

guard secrets. The medicine men and women must be protected from the curiosity of those who would misappropriate their knowledge.

Mistrust continues to reign. Recent history, that of European colonization, is at the basis of this attitude. Take for example, the age-old practice of the sweat lodge. The Finns have practiced it for centuries with their sauna; the Roman bathed in the *balneum*, immersed in small tubs of hot water, surrounded by clouds of vapor. The Aztecs thought that steam baths were a panacea against all ills. In Africa, sweating is both a therapy and a recipe for skin care. In Kenya, they keep hot embers under a dry clay seat. Women in particular enjoy sitting on it, surrounding themselves with a buffalo hide poncho reaching to the ground, as if inside a kind of private tent. Through sweating, they rid their skin of toxins and maintain its beauty.

Sweating, either in a dry or humid environment, was a universal practice from Japan to Russia, through northern climes as well as in old Mexico. However, while a Finnish proverb says that in the sauna one should behave as in a church, the American colonists found the sweat lodge sinful. Let's not forget that from the fifteenth to the seventeenth century Europeans considered bathing as unhygienic.

For pre-1850 Europeans, bathing and sinning were synonymous. In California, the Spanish Catholic missions forbade the use of the sunken oven which heated up the *temescal*. From that time, the Natives suffered from itches, tumors and other epidemics, for the heat not only cleans the skin, it also kills bacteria and viruses. Later, federal law prohibited dances and potlatches as late as the Second World War in the USA and up to the 1950s in Canada.

In addition, there is more to the sweat lodge than a physiological and psychological cleansing. The participation of the shaman in the sweating ceremony, through prophecies and the interpretation of dreams, enriched it with a religious and metaphysical experience.

I often heard Indian students in Omak mention that they had taken part in a sweat lodge ceremony. I became curious, as I did not comprehend the impact of this practice. I wanted to find out more. I had read here and there in creation tales that Coyote, besides creating the animals, had also given life to the sweat lodge, considered as a living being. I was quite intrigued.

The lodge seems to possess a transformative power. When it was covered with a bear's skin it took on, in some way, its initiatic properties, the hibernating bear being associated with both life and death. During its long slumber, the bear is close to the spiritual world. So, after some time in the darkness of the lodge the participant, like the bear, wakes up and returns to a renewed life. In case of serious illnesses, they might even escape from the claws of death. Many tribes, among them the Nez Perce, the Yakima, Okanogan and Colville, believe that the lodge is the body of the Great Spirit which gave names to the animals and took that form to come to the help of human beings.

One of the most widely spread stories among Salish Indians tells how the Great Chief or Great Spirit named the animals. The ever present Coyote played a major role, although he already had a name: *Sin-ka-lip,* the Imitator. But he preferred the name of a more powerful animal, like the Grizzly Bear or the Eagle. Concerned about her spouse's excessive ambition, Coyote's wife purposely forgot to wake him up. Sin-ka-lip went confidently, but late, to the Great Chief, who had stacked rings on a stick, each one bearing the name of an animal. By the time Coyote arrived, there was only one ring left.

Coyote asked to be Eagle, but the Great Chief told him that name had already been given out. Coyote then said, "I mean that I wanted to be Grizzly Bear." Too late! "Then, I will be salmon," added Coyote. But the Great Chief told him that all the names had been taken and that he would remain Sin-ka-lip. Then the Great Chief told him that his task, a difficult one, would be to help the human newcomers threatened by monsters trying to get rid of them. It will be Sin-ka-lip's mission to fight these monsters. Sometimes he will succumb, but Fox, his brother, will bring him back to life.

The Great Chief then decided to take on an available name, *Quil-sten,* meaning "He Who Warms up the Sweat Lodge." He announced, "Whoever wants to build me can seek my help."

The sweat lodge still plays an important role today. In the words of Spokane–Coeur d'Alene poet and novelist Sherman Alexie: "The sweat lodge is my church."

Sweat lodge ceremonies also play an important role with northern hunters. For them, the sweat serves not only to eliminate the body odors that the game might detect, but also to cleanse and purify the spirit and the soul. These sessions were of the utmost importance when hunting for bear, an animal most worthy of respect.

Chapter 41

Meeting the People of the Pacific Northwest

My days in Riverside were spent in meetings and encounters with a great many people. My documentary research was greatly facilitated by the assistance of Thelma, the charming Omak Community College librarian. R.C. Hoover, a professor of English, made valuable suggestions. I gradually found myself facing a massive accumulation of documents pertinent to the history of the Pacific Northwest and the culture of its various Native tribes. I understood that I was being encouraged to take a broad view of all the elements contributing to the geographical and cultural identity of the region.

From the perspective of mythological research, that kind of approach seemed appropriate. It would be an oversimplification to consider only the issue of the sasquatch at the exclusion of everything else. It would give it an undue importance and deprive it of the company of the main figures contributing to the imagination and the psyche of the Native North Americans.

The study of basket-making as it was practiced by the people living near the Columbia River could, if one wished, reflect practically all aspects of their civilization, for it touches on nearly all their activities. It mirrors the beauty of salmon or sturgeon fishing, as well as that of rodeos (a modern distraction); it carries simple and elegant geometrical designs.

Mary Dodds Schlick, a specialist in Pacific Northwest basketry, mentions only a single representation of sasquatch on baskets. It is much more common to find images of moose or birds.

Museums only rather belatedly appreciated the artistic wealth of baskets, bags, bonnets, hats and carpets. It is interesting to note that the explorers Lewis and Clark brought back from their western journey (1805–1806) a large cylindrical basket. Perhaps they realized, as Mary Dodds Schlick expressed it so clearly a century later, that basketry takes care of most of the problems of daily life.

I think of the myths of the North American Indians as a large carpet spread on the ground, the collective work of many weavers, working not far from each other. The patterns gradually take form. I also perceive them upside down, and it is difficult to understand them as they are still in the process of being created or modified. Further, these patterns often present a style that is either foreign to me or decidedly confusing. For example, among the Alsea Indians, a nearly extinct Oregon tribe, Coyote is at the same time a cultural hero and a divine prankster and transforms into Suku, a black bear. Suku then takes on the role of Transformer, bringing forth rivers and salmon and freeing the world from evil monsters. All the Alsea left to Coyote were its well known vices: lying, boasting, gluttony, obscenity.

There is no better way to show the essential duality of Coyote—as well as that of other major players like sasquatch and orca—than to consider the Coyote/Suku dichotomy. The twin-headed nature of the divine hero then becomes clearly evident.

I must add that the tapestry of myths developing under my eyes is the work of weavers working at different rhythms, even if everyone in a given region uses the same materials. Here, in the Pacific Northwest, bear grass, also called Indian basket grass, is widely used.

In this region, the bear is never far from sight. In late May 1994, in Okanogan, a town of 33,000, a yearling black bear was hit by a car at night.[1] The animal took refuge in a tree from which he was brought down with a tranquilizer dart. A few days later, a veterinary discovered a nonoperable fracture in its hip and the animal was euthanized. It was then buried carefully so that its corpse would not fall prey to scavengers.

In extremely cold winters, a grizzly may be seen on the mountainside, above Okanogan, an event that brings forth mixed feelings of fascination and fear. Incidentally, the Okanogan River was created by Coyote after he had done the Columbia.

The animal that most closely resembles man is the bear, walking on its hind legs, with a skeleton somewhat similar to ours. In the words of prehistorian Leroi-Gourhan, the bear is everywhere a man in disguise.

---

1. A year old black bear still weighs only about 45 pounds (20 kg).

An often told story in the Pacific Northwest is that of the Bear Woman. One fall day, a hunter discovered in the mountains an area where there grew edible plants. He decided to build a hut in that place so that his wife could gather wild vegetables. He then went hunting. In the afternoon, a grizzly bear abducted the wife, carrying her on his shoulder. The hunter told his father about the disappearance of his wife and of the presence of bear tracks in the vicinity. His father said, "Grizzly has married your wife."

The hunter could not sleep. After four days of insomnia, he finally fell asleep. His wife appeared to him in a dream and said, "Every morning, the grizzly takes me to dig the ground for edible roots. You will be able to kill him, but he is very strong and endowed with magic powers."

The wife then told the hunter how to prepare poisoned arrows and gave him the recipe for an ointment that would free her from the power of the bear. The hunter lay in ambush near Grizzly's den and shot him with two deadly arrows. He then freed his wife from the power of the grizzly by rubbing her with an ointment made of needles gathered from the top of a pine tree. However, his wife warned him: "The influence of Grizzly has not completely disappeared. You must build a special hut for me apart from the others. Your sisters will bring me food. In the spring I shall be completely yours again."

When the snows melted, she washed in hot water, then dived in the stream while her sisters-in-law rubbed her with fir boughs. The hunter also purified himself and they then lived together.

One can thus see how difficult it is to rid oneself of the influence of the bear. It is not easy for the wife to wash away her passage through animality.

A story told in the Amur basin, in eastern Siberia, speaks of a hunter who returned home to find his house empty. His sister had been abducted by a bear. Much later, the hunter saw her in a house deep in the forest. She explained that she had married and that her husband was away on a trip. The hunter left, saw a bear and killed it. He came back to offer the meat to his sister. Seeing the head and the fur, she recognized her husband and died of grief; her brother lost his mind.

The sexual power of the bear works on people of all countries, women being most strongly affected. Near Barcelonette, in the French Alps, the feast of Saint Ours (Holy Bear) is celebrated on

February 1st. The girls pray: "Give me a husband! Let him be as black as a chimney hook but make him a powerful male."

The return to wilderness through wild sexual attraction is the theme of many tales and even of some novels. One of these, Paul Doyles' *Nioka, Bride of Bigfoot,* describes Katherine, a young twentieth century American experimenting with New Age practices: yoga, vegetarianism, the I Ching, a visit to an ashram, a parade of lovers. She finds herself moved by the words of Chief Seattle:

> And when the last red man shall have perished and the memory of my tribe is but a myth among the white man, these shores will swarm with the invisible dead of my tribe.

Soon, Katherine falls in love with the ancient rainforest of the Pacific Northwest. She immerses herself into nature, going barefoot, bareheaded, wearing only a dress woven from cedar bark, in the fashion of the Indians of old. She actually looks like an Indian, and some people who see her speak of her as "the wild woman of the Olympic Mountains." She is now called Nioka. A photographer seeking to capture her image thinks of her as the counterpart of Ishi, the last Californian Indian.

Nioka makes friends with the animals of the forest: the moose, the beaver, but shies away from a relation with the bear. She speaks of sasquatch as a "sort of legend." Nioka seeks a fulfillment which she could not find in the civilized world. She works to attain it by weaving baskets, gathering berries, digging up edible tubers, listening to the song of the Olympic Peninsula's Hoh River.

Of course, *Nokia,* the novel, may appear somewhat ridiculous compared to traditional American Indian stories. The whites' efforts to reach sexual fulfillment, psychological equilibrium and a state of harmony with the universe sometimes seems foolish. It is as if, through adopting a particular fashion or under the impact of their disillusion with the consumer society, whites imagined that American Indian traditions would cure their ills and satisfy their yearnings.

I suppose that a storyteller such as the Okanogan Indian Harry Robinson would think that few whites could fathom the depth and subtlety of traditional stories, which he classified in two categories: those of the times when animals and people were not entirely separate, and those following that separation. The second period would encompass historic times, including a portentous date, that of Columbus' discovery of America.

Did the European newcomers ever understand the "power" that animals confer upon the Natives? Did they ever understand, as Harry Robinson tells it, how the grizzly bear, a female grizzly to be precise, which is by its nature half-animal and half-human, may be wild and ferocious, but also kind and helpful? The female bear has a lot in common with an Indian woman: she eats dry meat, camas roots, berries. Once again, the distinction between American Indians and the bear seems very faint.

Nevertheless, from the seashore to the Colombia plateau, Indians hunted black bears and grizzlies for their meat. They believed that if the bear died to feed them, so would the other animals. The success of the hunt for other game depended on the death of the bear. However the bear massacre perpetrated by Lewis and Clark horrified the Natives, who treated these animals with great respect. True, if a "bad" bear broke through a hut or stole from a food cache, it was tracked and killed. But only those individuals whose guardian spirit was that of a bear could go after the thief, for they were the ones who knew the "bear-hunting songs." It is important to satisfy the animal spirit in order to maintain the harmony between it and people, a harmony as fragile as dew on morning petals.

The Nez Perce held that a woman could not have a grizzly as a guardian spirit: it was too difficult to control and would bring ruin and unhappiness. However among the Kwakiutl, if one placed a bear's right paw on a young woman's right hand, she would gather many berries and dig many clams.

The wild side, violent and out of control, is represented in the Norse word describing northern European warriors: the raving mad *berserkers* (*ber* = bear, *serk* = skin), who fought either naked or wearing a bear's skin.

The shaman, wrapped in a bear skin, may also during a trance be seized by a fit of fury, which he can master to fight evil spirits. Indian hunters gave the bear in the animal world the same place that the shaman occupied in man's world. Just like the bear, that mysterious herbalist, the shaman, holds great healing powers, in part because of his ritual isolation. Solitude, like pain, opens the spirit to the world of the invisible. It is interesting to note that one often speaks of the bear or the sasquatch in veiled words; among many tribes from California to Alaska the bear is both feared and respected. To kill a bear with a club is the kind of contact which allows the hunter to "know" his victim in a way that cannot come about with arrows or spears.

To the hunt, I prefer reading tales of bears and sasquatches as a way to penetrate into the Indian mind and to meet a bear different from those we know. For example, this story told by Andy Joseph, whose guardian spirit is bigfoot:

> For many years I have gathered Indian medicine for my adoptive mother and other elders that I am related to. I have also gone after a few for my own use. [Here Andy described some of the medicines that have been used on him].
>
> The stinging nettles are good for, and also to prevent, arthritis and rheumatism. The first touch hurts, but it soon smooths out and feels like silk. Over all, the sweat house is the best healing place there is. You are next to your Mother Earth, you are as clean as one can get and your prayers are as strong as they can ever been.
>
> During my vision quest at age 11, I sweated for two weeks from a full moon at the end of May until a moonless night in June. I prayed for a strong spirit and to be brave because I chose rough land as my vision site. I also drank a lot of rose bush tea in the sweat house as a protector from an evil spirit...When I came out of my vision quest, I blacked out for a few hours and came to laying on my back with an old rock cup-like medicine bowl that had a red hot rock in it. My medicine man was dropping kinnikinnick into it and fanning the smoke into my face. It made me real sick and I heaved all over my face and body. I have never liked smoke since. I was dizzy-like and they gave me Indian tea so I could eat again...I believe that my Indian instinct will guide me to a needed medicine if need be...[I'm proud of Andy's trust, as reflected in the last phrase of his letter as follows] Thank you for giving me an opportunity to share a part of my Indian life with you.[2]

---

2. Personal letter from Andy Joseph, 1994. (Curator of the Colville Federated Tribes Museum), see Chapter 21.

Times have changed, the scoffing has ceased, and medicine men are now respectfully consulted as traditional healers the world over. Their teachings are now at the basis of a new science: ethnobotany. As for me, uninitiated to this lore, I nevertheless listen with interest to what the practicians and experts have to say. Quoting from my journal on June 7, 1994:

> I am speaking with Tillie George about medicinal plants, wild vegetables, places where were born the gods of local mythologies. I look at her, short, round, slanted laughing eyes, prominent cheekbones. She is at once slow and lively. With Indians, one must take one's time. She knows Vi Hilbert and Adeline Fredin.[3] She speaks with humor of story tellers; of the differences between the dialects.

At that time, I was at Omak Community College, attempting to learn some Salish from Tillie George, an Okanogan Indian. I had to listen carefully to hear her low, nearly monotone voice, just like the continuous bass tone produced by some instruments. She liked to speak of the plants familiar to her people, such as the camas, an edible bulb unearthed with a digging stick. She described the area where she was training students—mostly Indians—to dig the ground, on the shores of the Okanogan River, near a steep cliff. She outlined the silhouette of the cliff, sharp against the afternoon sky. The solid rocky background dominated her audience as a living wall, a curtain of quiet energy. The ground hid a tasty root, a promise of pleasure and health. Although sparing of words and gestures, Tillie's descriptions had a strong evocative power.

The elders made a deep impression on me. I asked myself repeatedly whether they actually behaved just so as to leave such an impression on their interviewers. I believe they carefully assessed the circumstances: Tillie George communicated her knowledge during lectures; Andy Joseph was speaking to me as a researcher. Their words were offered in

---

3. Vi Hilbert and Adeline Fredin are respected elders—keepers of cultural and religious oral traditional knowledge.

circumstances rising above the banality of everyday activities. They nevertheless retained an admirable simplicity.

Thelma, the most erudite librarian of Omak Community College, introduced me to an Indian of 50 years or so, whose words contrasted sharply with those of the above elders. Here's what my notes say about him:

> John has a feel for paradox, joking, comic gestures; he is proud and a bit of a cynic. That makes him very interesting. He suggests that I should not overly romanticize the Indians' life. In his view, there never existed an idyllic past before the conquest by the white men, followed by a dark postconquest period. Too much independence or originality annoyed the chief. Opinions could be expressed during tribal council, but minority views were not listened to. The dissenter was thrown out and might be forced to marry outside the tribe, for example. However John trusts me enough to suggest that I should visit Adeline Fredin who has an important official position within the Colville reservation.

Slightly confused, I return home, or at least to my home in Riverside, Ed's old school. Everyday life—so called ordinary reality—is full of perplexing circumstances.

Ed has just had enormous billboards painted to attract the attention of people driving up Highway 97, an hour away from Canada. Canadians do come from time to time to buy gold prospecting supplies from his store.

Tom, the artist, usually earns a living customizing and detailing cars and trucks. He lives with his wife and their three children in a converted bus. He travels to automobile shows and all kinds of vehicle exhibits, looking for clients. Tom and his wife are home schooling their kids. They say they are doing a better job than the school system. The children seem alert, intelligent and polite.

On one of the billboards Tom showed a prospector, his panning dish in hand, leaning over a woodland stream bordered by tall Douglas fir under a bright blue sky. This bright and simple scene will decorate one of the walls of the old school where Ed now lives.

In the morning, Ed gives panning lessons to Tom's family at a metal tub raised over four posts. With a waist-high tub one can practice panning without having to crouch. In the tub, there is water, sand and some fake gold flecks. Ed plans to barter an introduction to prospecting for part of Tom's painting bill.

In the evening, I return to Riverside, driving along the Okanogan River, flowing between low hills. Horses stand out in the pastures in the golden afternoon light.

Ed has decided to do a slide show on geology, focusing on gold deposits. This is the theoretical counterpart of the practical lessons offered to Tom and his family. A lively and gifted teacher, Ed keeps his audience enthralled.

By the time we cook dinner—fried chicken and onions—it is already dark.

The following day, at the Community College, a young student approaches me to ask my views about the street curfew recently imposed in Wenatchee (22,000 inhabitants) and Electric City (less than 1,000 inhabitants). This last village is near Grand Coulee and Coulee Dam, the three villages counting up to about 3,000 people. Andy Joseph lives in Coulee Dam. How could he possibly have met his guardian spirit if he had been locked into his room after sunset? I answer that it would have been outrageous to lock up Joan of Arc or Arthur Rimbaud.

What is happening in towns and cities, great and small, is far from conducive to initiation practices. Ten years later, similar measures were to be adopted in France to control rebellious youth.

I tell Thelma, the Community College librarian, that Adeline Frelin has granted me an interview at her office in Nespelem, at the heart of the Indian reservation. To reach it, I must drive through the reservation on a winding road, climbing through the forest. Halfway there the road narrows and passes between two cliffs. Thelma tells me that at that very spot the temperature drops abruptly. It is also said that there is a time warp between Omak and Nespelem, so that it is impossible to estimate a travel time between the two towns. Thelma says that she is always afraid of arriving late when she takes the Nespelem road. She seems quite serious about this. Actually, I had already the impression that Nespelem was quite far.

Although I suspected that John might have taken a malicious pleasure in teasing me a little, I could not fairly accuse Thelma of such an intent. Nevertheless, without connivance, John and Thelma both seemed working towards the same goal: to make my interview with Adeline Fredin most awkward.

If the way to Nespelem seemed such an ordeal, it is most probably because I could remember my winter trip with Ed, in the snow storm. If I manage to cross the forest again without a problem, will I have to face a dragon, a tiger or an ogress?

Between classes, students and professors sit at wooden tables on the lawn of the Community College. People chat, smoking and drinking coffee. Carol McMillan, a professor of anthropology, tells me that there always reigns around Omak and its neighborhood a mysterious climate, as if at the junction of two planes of existence.

One day, Carol introduces me to a few students in the hall. She briefly explains to them the gist of my research. Greg, a Nez Perce student, says he would be willing to introduce me to an elder who knows a lot of sasquatch stories. The young woman next to him, a Colville Indian, reacts quickly:

> You shouldn't reveal too much to these cultural anthropologists who come here to steal our soul. We are in fashion again. This happens every 15 years or so!

After a moment's thought, I withdraw from the conversation, without however feeling any shame about my investigation. I know I might be wrong, but I listen carefully to the advice I am given.

Not long after this, the same young woman comes and sits next to me. She is beginning legal studies: she wants to become a lawyer. In turn, I gradually tell her about my plans: I will soon go to the University of Idaho, in Boise. I will stop in Coeur d'Alene and Lewiston on the way there and return via the Lewis and Clark trail.

She then tells me most seriously that for her Lewis and Clark are for ever linked with this region, northern Idaho and eastern Washington. After their expedition, the Far West was no longer inaccessible. It would quickly change under the impact of the invaders imposing their views on the Natives. The path of the conquerors remains invisible to the naked eye: they left no monuments, palaces, castles, forts, fortified bridges, statues. All that is left are the accounts of the explorers and the maps showing their itinerary, well known to Americans who learn about them from their history teachers at school.

The images seen in the lower row in this Yokut native basket depict the "hairy man," now believed to be the sasquatch.

One must rely on imagination since there are no monuments to commemorate the explorers' path. The vastness of the scenery and the menacing wilderness conferred to the region a mythical aura, both fascinating and sometimes terrifying for the American expedition.

The intensity of the young Indian woman led me to think about the historical intrusion of Lewis and Clark into the Pacific Northwest. All my travels in that direction were from now on to take on a new perspective.

Chapter 42

The Conquest: In the Shadow of Lewis and Clark

Just beyond the small wooden bridge next to the sawmill one enters Indian territory, the Colville reservation, so named after the Hudson's Bay Company's Fort Colville, established in 1825 at Kettle Falls, on the upper Columbia, near the Canadian border.

The Colville Indians included the Okanogan, the Sanpoil, and the Sinixt, are all speakers of Interior Salish, so called to distinguish it from Coast Salish, spoken by the tribes of the Pacific coast. "Kettle Falls was early fixed upon as the site of an important post by the Hudson Bay Company and brought with it usual advantages and disadvantages of white contact."[1]

According to Lewis and Clark, the Colville tribes numbered 2,500 in 1780; in 1937, their numbers had fallen to 322. On President Thomas Jefferson's orders, William Clark and Meriwether Lewis undertook an expedition to the Pacific Northwest, which made them famous. Starting on August 3, 1803, they immediately began a journal which ran up to 5,000 printed pages. As my friend Lauric Guillaud put it: "Before conquering the West, one had to write about it."[2]

The explorers started near St. Louis, travelling up the Missouri River. In September 1805, they crossed Lolo Pass, southwest of Missoula, Montana, and reached what is today Idaho, where they soon met the Nez Perce Indians and their neighbors the Coeur d'Alene and the Sanpoil. Curious about the intruders, the Natives quickly checked out the expedition. Three boys approached a group led by Clark. Frightened by the appearance of the *suyapos,* as they called the strangely bearded explorers, they hid in the bushes. The first white man met by the northwest Indians was bald and bearded: they named him *suyapo*: upside-down head. Approaching unarmed, Clark gained the confidence of the young Natives who led him to

---

1. John Reed Swanton, Indian Tribes of Washington, Oregon and Idaho, p. 24
2. Lauric Guillaud, La Terreur et le Sacré, p. 129.

their village. The chief, Broken Arm, was away on the war path, fighting the Shoshone (an expedition from which he was later to return with 42 Shoshone scalps). The Nez Perce were friendly towards the explorers, even though they were guided by a Shoshone woman with babe in arms, Sacagawea. The women and the elders of the village prepared a feast of buffalo meat, salmon and camas roots.

In spite of some difficulties in communication with the Sahaptin-speaking Nez Perce, the explorers followed local techniques to built five boats in three days, burning the core of cedar trunks to dig out canoes.

A few Nez Perce joined the expedition, heading downstream to the Pacific coast. The Indian guides were most helpful in avoiding waterfalls and sandbars; they held the suyapos' lives in their hands.[3]

The terrifying rapids of the aptly named Snake River carried them into a strange, desolate and inhospitable world of emptiness. The descent of the Columbia took place in the most complete bewilderment.[4]

The wilderness is full of terrifying creatures. Bears haunt human consciousness. "Repeated attacks created among the men a state of mental chaos."[5] The black bears and the enormous grizzlies are loaded with terrifying mythical potential. Like the sasquatch, they appear practically indestructible. However, while the latter is depicted as a peaceful creature, the bear, guardian of the wilderness, bristles with indomitable ferocity.

The snakes, particularly the rattlesnakes, worried the travelers. Reaching the plain, 50 miles (80 km) west of Missoula, the climate becomes progressively drier. As far as the Cascade Mountains, the scenery is desertic, a hot environment favored by rattlesnakes, as well as by black widow spiders. Large—up to eight feet (240 cm)

---

3. Today, practical jokers love to send naïve visitors professing an interest in ethnology on a search for traces and artifacts of the Suyapo tribe along the Columbia River.
4. Laurie Guillaud, op. cit. p. 153. The Shoshone were sometimes called the People of the Snake by the French.
5. Lauric Guillaud, op. cit. p. 143.

long—and inoffensive, but frightening bull snakes, easily confused with rattlesnakes, also live in the area.

Lewis and Clark describe an encounter with rattlesnakes: "A short distance above the mouth of this creek are several curious paintings and carvings in the projecting rock of limestone, inlaid with white, red, and blue flint of a very good quality. The Indians have taken off this flint great quantities. We landed at this inscription and found it to be a den of rattlesnakes. We had not landed three minutes before three very large snakes were observed in the crevices of the rocks and killed." (7 June 1804)[6]

Today one may from the comfort of one's car follow the Lewis and Clark Trail, starting in Lewiston, Idaho, crossing the Snake River to Clarkston, Washington and continuing west on US Highway 12. On a sunny day, nothing can bring to mind the nightmarish conditions faced by the expedition, which finally wintered at the mouth of the Columbia with the Clatsop Indians. Clatsop means "dried salmon" in the Chinook jargon. Lewis and Clark estimated their numbers at 300; in the 1910 census, there were only 26 left.

Lewis and Clark arrived in the Nez Perce territory at the beginning of fall, or *Pekhoonmaikahl,* according to the designation of the seasons used by the Natives. The trees were beginning to yellow; women and children were collecting berries for drying. The salmon were swimming up river and were being caught with hooks, nets, dipnets and harpoons.

In the words of Clifford Trafzer:

> For the Nez Perce, catching and eating salmon was a spiritual activity, part of their religion. Their spiritual beliefs still require that they eat salmon and drink water in communion with the Creator each year.[7]

---

6. The Journals of Lewis and Clark, National Geographic, Adventure Classics, Washington, DC, 2002, p. 31.

7. Clifford E. Trafzer, The Nez Percé, p. 21. Trafzer is a member of the Wyandot or Huron tribe, formerly living in southern Ontario. They were scattered by their relatives the Iroquois. From a base in northern Ohio, they fought on the side of the British in the American Revolution. In 1843, they were displaced by the American government to a reserve in Kansas.

Of course, the tribes of the Pacific coast also show great respect for the salmon. They cure its skin to make a pliable leather; they also cut its flesh and smoke it into lox, which can be preserved for a long time. A Kwakiutl poem celebrates the salmon:

> I am salmon,
> Ever returning swimmer,
> I am salmon,
> Ever reborn swimmer,
> I am salmon,
> Swimmer ever reappearing,
> With the night's star
> Fish of the moon.

Believe me, feasting on the salmon is the occasion for the most joyous festivities.

The Nez Perce also hunted deer, bears and moose to stock the winter larder. The fall was a joyful period, sometimes blending with some seriousness. This was the time of year when young boys and girls, 10 to 11 years old, left on the quest for their guardian spirit.

During their visit with the Nez Perce, the white explorers enjoyed their friendly hospitality and the abundant food they offered. They benefited from their knowledge. Preparing their departure for the Pacific and the final leg of their journey, they were rather anxious, again facing the unknown. The Nez Perce had told them about Coyote, the divine trickster, fully living up to his reputation of supreme cunning as the creator of their tribe.

Free and easy, at the dawn of time, Coyote met an enormous and insatiable monster, *Iltswewitsix*. As the monster had seen him, Coyote could not refuse its challenge. "Take a deep breath and eat me if you can," said the monster.

Coyote only managed to eat one of the monster's legs. Iltswewitsix then got ready to gobble him up. "Wait," said Coyote, "let me go into your stomach so that I don't have to suffer."

Stupidly, the monster agreed. Once inside, Coyote lit a fire, and with the five knives he had hidden, cut out the monster's heart. He then cut out the rest of the body in pieces, which he distributed to each of the groups of humans around him. Finally, he noticed there was not a single piece left for the Nez Perce. That's when he decided to create them using the monster's blood, a source of courage, strength and honor.

On their way to the Pacific, Lewis and Clark traveled along a hillside near the Clearwater River—the heart of the Kamiah monster (today, Kamiah is a small Idaho town). That, it is said, is the birthplace of the Nez Perce and is venerated as such.

The expedition endured a difficult winter at Fort Clatsop, while the Nez Perce, cozy in their longhouses counted on their accumulated provisions. As the snow lasted longer than usual, they had to do some hunting to renew supplies—with little success. In any case, as in previous years, the tribe survived the winter, engaging in a series of ceremonial dances and storytelling, weaving baskets, making clothes, repairing fishing and hunting gear, all the while perpetuating binding tribal lore.

Finally came the season of blooming flowers, *Lahtetahl*. Lewis and Clark started back on their journey home. They stopped at the villages of chiefs Broken Arm and Twisted Hair, among others, and took back their saddles, gun flints, and powder left for safekeeping. In spite of the severe cold, the Indians had taken good care of the Americans' horses. The expedition waited until mid-June, for the snow to melt off the Bitterroot Mountains, when they took leave of their Nez Perce friends.

By the end of June they were resting their tired limbs in the hot pools of Lolo Springs before facing the steep slopes of the Rockies. "Lewis was already thinking of a time when white families would settle on the lands having been the exclusive domain of the Indians, a situation that the latter could hardly imagine."[8]

---

8. Trafzer, op. cit. p. 21.

Chapter 43

The Indians Dispossessed

For a few years, the Indians continued to live at the rhythm of the seasons, making modest demands on the resources of their environment. But soon news spread that two prospectors had found gold in the Clearwater River, in the heart of Nez Perce territory. While the whites rejoiced, the Indians were dismayed.

In 1863, the whites held a council at Fort Lapwai. They informed the Indians that the boundaries of their territory, as defined by the Walla Walla treaty of 1855, would be drawn in to encompass only one-tenth of its former area. For many Nez Perce, including an influential young leader, Chief Joseph, this was the year of the "Thief Treaty." Joseph's father had so named him in memory of the cordial relations he had maintained with the trappers and traders who had followed in Lewis and Clark's footsteps.[1]

According to Chief Joseph, "the Creative Power, when He made the earth, made no marks, no lines of division or separation on it. The earth was his mother and he was made of the earth and grew up on its bosom."[2]

In 1877, General Howard gave the Nez Perce a month to settle on their new, smaller reservation. The Indians gathered at the Field of Camas, a traditional meeting place opening on a spectacular panorama, a symbol of the freedom they were soon to lose.

Unfortunately, three young braves, driven by a wish for vengeance, killed three white men and wounded another. They were soon joined by other warriors. After some confusion, the Indians decided to leave for Montana and then cross the Canadian border.

---

1. Chief Joseph of course also had an Indian name: Hin-mah-too-yah-lat-kekt, or Thunder Rolling Down the Mountain.

2. The Nez Percé, op. cit., p. 69.

They followed the path taken by Lewis and Clark on their return voyage, but under tragic circumstances, with Howard and his troops on their tail. After two months of flight, about 800 Nez Perce men, women and children reached Big Hole, Montana, where they set up camp, thinking themselves in relative security.

Colonel Gibbon and his men attacked at dawn on August 9, 1877. The officer did not want to take any prisoners. There were soon 60 to 90 victims, mostly women and children, but the warriors continued to resist and Chief Joseph managed to gather the survivors to lead them across the border to Canada, where they would find peace and rest with Chief Sitting Bull, who had already taken refuge there.

Chief Joseph

Pressed by the US Army, and against the advice of Chief Joseph, the Nez Perce rested after crossing the Missouri River at the foot of the Bear Paw Mountains. The Canadian border was only two days' march away, about 40 miles (65 km).

Alas, the soldiers cut them off and attacked their camp. Men, women and children fought back vigorously. Colonel Miles ordered the camp besieged. In the snow, suffering from cold and hunger, the 400 survivors would have fought on had Chief Joseph not ordered the surrender on October 1, 1877. Miles and Howard promised to escort the Indians back to their reservation in Idaho.

However, William Tecumseh Sherman, Commanding General of the US Army, had just declared: "For the more I see of these Indians, the more convinced I am that they all have to be killed or maintained as a species of paupers."[3]

The terms of the surrender were not respected. In July 1878,

---

3. The Nez Percé, op. cit. p. 89.

after eight months in captivity, the Nez Perce were sent to Kansas. Twenty-one had died under the hardships of internment. Disease took many more. Eight years later, there were only 268 survivors. Finally, the government returned them to the Washington Territory, splitting them in two groups: Chief Joseph and 150 Indians settled on the Colville reservation; the others returned to the Idaho Nez Perce reservation.

> **ASIDE:** While describing these historical events, I cannot help remembering that I was not so long ago confronted with pictures of a tragedy which occurred in Paris on October 17, 1961. In 1962 I was a teaching assistant in a grammar school in Glasgow, Scotland. One evening, as I came out of a cinema, I was handed a leaflet which left me awestruck: it showed demonstrators beaten to death by the French police. Our president was General De Gaulle and I could not believe my eyes. The peaceful demonstrators were Algerians asking for the independence of their country. There were between 32 and 325 victims (the numbers, actually quite high, varied widely according to sources: government, political parties, media).The only visual evidence was provided by French photographer Elie Kagan and also a crew of American journalists who filmed the event and helped some of the wounded. One would have to wait until the mid-seventies before the media mentioned the brutal repression. But I knew before most of my fellow citizens and felt most ashamed, especially as a Frenchman in the UK.
>
> Another demonstration (February 8, 1962) against the war in Algeria claimed nine victims in Paris. The ceasefire of March 18, 1962 marked the end of the Algerian conflict. My pre-military training was not to be put to actual use then.
>
> I remain proud of my country, but I deeply regret the brutal measures taken by our statesmen.
>
> Civilized countries are never immune to violence and injustice.

The long trek to which Chief Joseph's and his tribe were subjected is typical of the fate of the Natives. Today, Chief Joseph's reputation still lives among the Nez Perce. His qualities as a leader, diplomat and strategist, his great humility, his commitment to his people, and the tenacity with which he pleaded with the authorities in their favor, are widely admired on the Colville reservation and among American Indians in general.

Should you one day pass through Nespelem, you will have no difficulty finding his grave in the small cemetery. Please pause and honor his memory.

The last great conflict with the Indians took place at Wounded Knee, South Dakota, a few days after Christmas 1890. It was triggered by the death of Sitting Bull, killed by the officers who had come to arrest him. In the end, 300 Indians were killed, two-thirds of them women and children. There were 49 victims among government forces. After that, there were only a few minor and short-lived skirmishes.

It is interesting to note that locations where some battles took place include the like of Bigfoot Pass, Bigfoot Creek and Bigfoot Hill. In a short anonymous article in *The Track Record*,[4] an American Indian wrote that those places had a particular spiritual signification, related to the flight of the tribes pursued by Colonel Miles. Bigfoot is the English translation of *Bekaycho*, from the language of the Sioux, the Crow and the Chippewa.

In the words of the anonymous author:

> after the bloody massacre at Wounded Knee, the Shaman leading the band of Sioux followed the tracks of this giant protector North to safety, he/it (Bekaycho/Bigfoot) always showing the way but rarely seen. If he/it was seen, bad luck would follow (hunger, missing children and the like) and if the trail was abandoned certain catastrophe would ensue.[5]

The same author concludes with a reference to Star Wars figure

---

4. The Track Record, No. 103, January 2001, p 7.
5. See Appendix 2.

Chewbacca, possibly inspired by the mythical Bekaycho, helper of the oppressed.

It would appear that the fugitives had fled northwards under the direction of a shaman. Neither white nor Indian historians mention this. Earlier, Chief Joseph, war leader, wise man and visionary had succeeded in avoiding much bloodshed in leading his tribe towards Canada. But in December 1890, the fleeing Sioux, led by Chief Bigfoot, a historical figure, aimed for the Pine Ridge reservation, in South Dakota. Chief Bigfoot had decided to yield to the demands of the American authorities, but a warrior excited by the war-like harangues of medicine man Yellow Bird fired on the soldiers. In the following shootout, Bigfoot was killed.[6]

The erroneous account of the events at Wounded Knee, as alleged by the same anonymous writer, is revealing from a mythological point of view. We find in it nostalgia for the old days, for a world where the fate of each tribe was in the hands of a wise chief who was at the same time a strategist, a prophet, and a real guide of his people, untainted by the white man's values.

Beychako/bigfoot was the very embodiment of such legendary qualities.

---

6. See James Mooney, The Ghost Dance Religion and Wounded Knee.

Chapter 44

Myths: An Endless Tapestry

Taking a close look at the tapestry woven by the creators of mythical figures, we can perceive a certain degree of unity in the depiction of the great myths. At any one time on the planet, the same ideas move people everywhere: this becomes more noticeable when taking some distance. Sometimes however, upon closer inspection, it appears that some groups remain inspired by one or more myths which are clearly outdated. Myths, like living beings, are born and eventually disappear—except perhaps in the case where a myth survives against all expectations.

We all have memories of the history of great religions. Take Abraham, for example, submitted by the voice of God to a demanding test, which he obeyed, readying to sacrifice his son Isaac. Luckily, a ram emerged from the bushes and offered itself instead (around 2030 BC). In Pichon's words:

> The ram is a Chaldean symbol; it belongs to the zodiacal bestiary where the sun-god was represented by the Lion, knowledge by the Serpent, Kinship by the Bird, and Creation by the Bull.[1]

The ram, god of justice, will be served by an inspired man, a scion of the royal family of Thebes: Moses, priest of Osiris. Banned from the symbolism that inspires his teaching are the virgin, the fish, the bull (in the form of the golden calf) and gemini (the twins). Pichon continues:

> However, the new god, the living god, is far from showing up as homogeneous. Rather than possessing a well defined appearance, it is still only a tendency, likely to adopt almost any form, a tendency which must also eventually fulfill all

---

1 Jean-Charles Pichon, *Histoire des Mythes*, p. 75.

aspirations, all of humanity's dreams. It emerges only with difficulty from a pagan world reluctant to accept the unification of faiths.[2]

It is only 2,150 years later, after a long gestation, that the new faith emerges, personified by Jesus of Nazareth. The Kabbalists of the first centuries of the Christian era claimed that Jesus was a reincarnation of Esau, the long-haired being, Jacob's twin brother. The argument is based in part on the similarity of their names, but also on the contrast between the Christian and the Jewish faiths. Like the Christian church, which opens its doors to all, Esau had opened the race of the Patriarchs to foreign wives.

The gemini resurface in the cult of Remus and Romulus in Rome and in the veneration of the Greek Dioscuri, Castor and Pollux. In addition, the new religion brings in the virgin: Mithra, like Krishna, was born in a cave from a virgin mother. Through baptism, whether in the Jordan or the Ganges, the Essenians and the Buddhists welcome the new god announced by Job's leviathan, Tobias's fish, Jonah's whale. Jesus's sign is to be the fish.

Before it emerged fully, Christianity went through a prolonged gestation. It also had a number of serious competitors; the god Mithra, for example, associated with the god Ahura-Mazda and its sixth century BC prophet, Zarathustra. Mithra almost outcompeted Jesus. The Roman legions carried its cult to all corners of the Empire. The emperor Nero was a convert, submitting himself to the ritual anointment with the blood of a bull. One of my students, an Iranian, told me that her father had converted to Mithraism in reaction to the Islamic fundamentalism of Ayatollah Khomeiny. There are also adepts of Mithra in India in the form of the Parsi religious minority. Their habit of exposing the bodies of their dead on "towers of silence" where the vultures devour them is currently being discouraged by the authorities for public health reasons.

On display at the Périgueux Museum, there is a beautiful taurobolic altar.[3] A few kilometers away, at Aubeterre sur Dronne

---

2. Pichon, loc. cit. p. 76.
3. Turcan, R and A.Nevill, 1997, p.60. An excellent reference to the religions of the Roman Empire and the artifacts that document them.

(Charente department, France) a church, carved into the interior of a cliff surmounted by a medieval castle, shows traces of a succession of faiths, successively harboring prehistoric tombs, a Mithraic temple, later converted into a Christian church, and a Catholic mausoleum. In its early days, Christianity had to compete with many other sects, in particular the gnostics. One of these, the Ophites (second and third century AD) worshiped the serpent, as the bearer of sacred knowledge. The Greek *gnosis* means knowledge; ophite sounds a little like *sophia,* Greek for wisdom.

Today, worshippers of the serpent are still found in many countries of Africa, Asia and America. In some states of the USA, particularly Kentucky and Tennessee, the faithful handle vipers to prove their faith. George Went Hensley, the initiator of this practice wanted to show that a true believer could handle a venomous snake barehanded without any danger. The feelings of exaltation and superiority arising from such an achievement illustrate in a way the limitations of the puritan faith preached by the fundamentalist religions of the South. Hensley died of a snake bite in 1955. Africa remains one of the areas where the cult of the serpent still plays an important role.[4]

One sometimes gets the impression that some people became fixated on a particular mythology and that time has since stood still in matters of faith. There survive a great many practices, vestiges from ancient beliefs: worship of the gemini, nostalgia for the Earth Mother, the survival of Jewish legends in Black Africa, such as the sacrifice of the ram, and a variety of domestic and patriarchal customs.

As one may witness by traveling around world the many stages of the progress of civilization, one may similarly observe the development of religious history by going from one country to another. One expects to meet astronauts at Cape Kennedy and hunter-gatherers in Amazonia or Papua New Guinea. That their beliefs systems may differ should not be a matter of great surprise. They all express a hope, an expectation to some degree or other, the moments of fulfillment being rather few and far between. Local beliefs are instruc-

---

4. See Appendix 3.

tive as living witnesses of a long-gone past, or as reflections of archetypes still influential in today's world.

When the local gods are put down by invaders, as was the case in America, one could speak, as did Jean-Charles Pichon, of an "abandoned people." But, Pichon quickly adds that people cannot live long without faith. Conquered people secretly maintain their rites. Sometimes they borrow fragments of their conquerors' religion. Bringing together scattered beliefs, they end up with an ever-evolving syncretism. For example, the Peyote rituals originating in Mexico and spreading in the southern plains in the nineteenth century, found formal expression in the Native American Church in 1918. This Native church is represented in nearly all US Indian reservations. It brings to the Indian communities a foundation of social and moral stability—often severely lacking among the youth.

The Sun Dance ceremony among the Natives of the plains and the Columbia Plateau also draws participants from other regions. It includes sweating, sometimes flagellation, and exhausting vision-provoking dancing. The buffalo plays an important role during the many days of the ceremony. Trials, sometimes painful, emphasize the initiatic nature of the Sun Dance, which invites novices to grow in wisdom over successive years.

I have already mentioned intertribal borrowings. In the past, there were already numerous contacts through commerce, a semi-nomadic lifestyle, and socio-religious exchanges. Contacts increased markedly in the twentieth century thanks to improving means of communication.

A baffling number of entities crowd the North American Indian mythological pantheon. Some groups cover a single basic essence, for example, the divine trickster is usually Coyote, but sometimes Hare, and more rarely Raven or Raccoon, etc. Others play secondary roles, for example, Wind Woman, a figure limited to the Alsea tribe of Oregon, or Porcupine, often appearing in Coyote's adventures. Fortunately, some traits are enduring: the buffalo, symbol of the sun and of creation, reminds one of the bull, symbol of Mithra.

Among the plethora of figures appearing as masks, ornaments of various kinds, songs, dances and stories, one should try to identify the major figures of the American Indian pantheon. The basic question, the same question that arises when studying the ancient as well as the modern civilizations of Europe, India, China, Egypt, etc., may be expressed as follows: What underlying principle is suggested by the ensemble of these figures? Is there on this vast tapestry of myths one particular silhouette that stands out?

The hardships endured by the Indians were rarely noted at the time; they found few defenders among the whites. An exceptional testimony was that of Alfred Downing, a surveyor who traveled with an Indian crew on a rowboat down the Columbia from Kettle Falls to the mouth of the Snake River. An adventurous 400-mile (600-km) trip—the Columbia, still unharnessed at the time—offered challenging surprises, like the Nespelem Rapids. Downing found the Indian rowers competent, energetic and good company. He regretted how the United States had deceived the Indians, ignoring one after the other the treaties signed with the various tribes. Downing's concluded: "The step from barbarism to civilization is a long and slow one; but from civilization to barbarism is a quick and easy one, to my mind."[5]

Standing in front of Chief Joseph's modest tombstone, in Nespelem, one might perhaps dream of a different destiny for the Indians. At the time of the exodus with Chief Joseph they were entering a very gloomy period. Except for a few *coureurs de bois* and anthropologists, like Franz Boas and James Teit, the colonizers scoffed at their gods, and over the following century many of the Native people even forgot about them. However, some of these deities continued to exert their influence, while others are still slumbering, biding their time. Again one might ask—can one discern a dominant silhouette emerging from the mists of mythology?

---

5. Alfred Downing, *The Region of the Upper Columbia River and How I Saw It*, p. 3.

Chapter 45

On the Way to Nespelem

Commenting on Lewis and Clark's westward progress, Lauric Guillaud suggests:

> The march to the West is associated with a voyage of initiation, recapitulating the ancient myths. Lewis, like Jason leading the Argonauts, sees the Symplegades[1] crashing together to block the path of the uninitiated. It is not by chance that he writes: "The striking appearance of this place led me to call it the Rocky Mountain Gates."[2]

Just like Jason and Lewis, I worry about the passage on which depends the success of my visit. The Stone Gates, or the Time Wardens, to give them a name, have nothing in common with the Symplegades, the moving rocks which almost crushed Jason's ship, *Argo*. Besides, my trip is quite short, 40 miles (65 km) or so, and I'm driving a rather prosaic vehicle—a small Japanese compact.

Nevertheless, how can I forget the area's geological past? I can clearly recall the enormous rocks which strew the slopes leading to the Columbia. These boulders were moved and fashioned by ice and water in a natural turmoil reflected in the Danish etymology of their name, *buldre,* to roar like thunder.

Just recently, an old Indian, Charlie, was telling me about clashing boulders, in a story about Coyote's intervention in a dispute between two brothers. It is precisely by smashing boulders together that Coyote made such a clangor that the brothers stopped fighting.

---

1. The Symplegades: mobile crashing rocks at the mouth of the Bosphorus, threatening Jason's quest for the Golden Fleece.

2. Lauric Guillaud, *La Terreur et le Sacré*, p. 145.

Flakes shot off the crashing rocks; Coyote picked them up. He had just invented arrowheads.

What else did he not invent through a succession of bizarre errors, coarse humor, and even scatological jokes! Mistakes breed novelty; laughter is fertile ground for wisdom. "Coyote is a master. There are others. Find your own."[3]

It is up to us to pick those rock chips that will turn out to be useful. Let's choose wisely. For example, let's not pick up old arrowheads. If they lie on the ground it's because they missed their goal. One should let those that were missed live in peace. They deserve to survive.

Often, Coyote's adventures provoke laughter. Once a week, I went to the Omak Community Center for lunch with Indian elders and I listened to their stories. Stories about Coyote often conclude with guffaws of laughter. At first, I am a little taken aback, handicapped and somewhat set apart by my serious attitude. Gradually, I learned to listen to the elders with respect.

Charlie is an elder who loves to laugh and provoke laughter. Once, he told the story of how Coyote dismembered the body of a cannibalistic monster that used to live at the mouth of the Palouse River.[4] The cannibal was claiming many victims and Coyote decided to rid the area of this curse. After a long and bitter fight, Coyote emerged the winner. He cut up the corpse of his enemy and threw its head in the river towards the Wishram tribe; from then on, they had large heads. He threw its scalp in the water for the Crow, who now have long hair. He cut out the chest and the ribs for the Nez Perce, who became stout. The legs, he also threw into the water for the Blackfeet, who are known for their long legs. The heart he kept for the Coeur d'Alene, courageous but sometimes cruel warriors. He forgot no one, taking particularly good care of the Salish. When he was done, there was only one piece left—the penis. Charlie claimed

---

3. Gérard Delfe. *Le Dieu Coyote*, p. 173.

4. The Palouse River is an affluent of the Snake River, which it joins near Starbuck, Washington. It runs between the Coeur D'Alene reservation on the north and the Nez Percé reservation on its southern side (in Idaho), passing through the villages of Potlatch, Idaho and Palouse, Washington.

that he had been fishing nearby and that he became the beneficiary of the enormous organ!

There is no doubt about Coyote's obscenity, or his maliciousness. When he was creating rivers, he wandered near the Colville reservation, in Nespelem. Somewhat tired of his lengthy voyages, Coyote planned to stay for a while in the area and to take a wife. The Colville Indians refused his request. Angry, Coyote decreed that the salmon would no longer swim as far upstream as Nespelem.

Was Coyote a bit of a prophet? Salmon were soon to disappear from northeastern Washington State. The series of dams on the Columbia made their upriver migration nearly impossible. Remaining accessible spawning areas are found further south, near the river's mouth. Today, salmon numbers are also diminishing because of water pollution by a variety of chemicals.

Bigfoot hunters are not easily discouraged. They believe that the hairy giants, like the bears, feed on migrating salmon. That's why during the spawning season, from August to October, the shores of the Columbia and of its affluents near the Bonneville Dam are suitable search areas for sasquatch investigators.

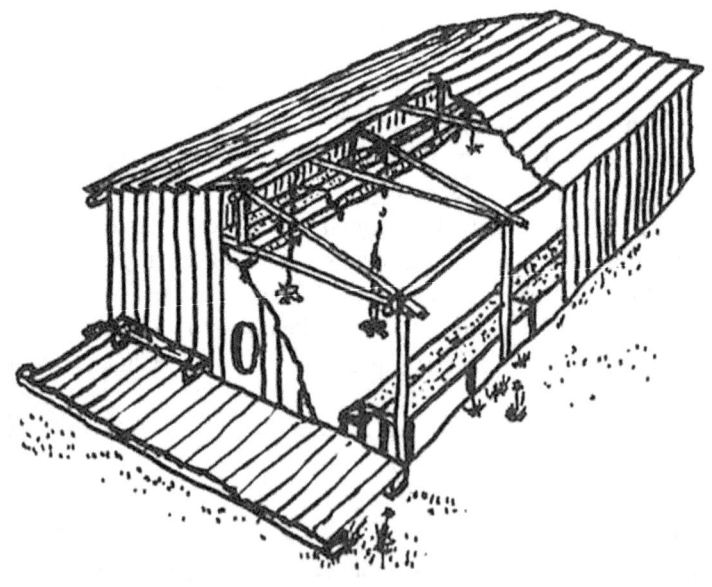

Communal dwelling, the longhouse.

Chapter 46
Vi Hilbert

The story of Coyote-creator-of-rivers was collected by Vi Hilbert[1] in 1982 from her cousin Walt Williams. One must give due credit: a story lives only through the person who tells it. Through association, the storyteller shares the existence of Coyote, Fox or Wolf. A collection of stories of the kind gathered by Vi Hilbert covers many generations. That is also the case when Walt Williams and his kin gather during the winter to tell a round of stories in the longhouse he built in Nooksack territory.

Many different dialects are spoken by the Indians in this part of Washington State (Skagit County, north of Seattle). Vi Hilbert is fluent in Lushootseed, a phonetically complex language. A simple word, *haboo* (stress on *boo*) deserves mention: it is used to encourage the storyteller from time to time; it means, "continue!"

The sometimes very lengthy legends speak of the Age of Myth, before the transformation of the world: "long ago...before the change, there were no humans as we know them. All of the beings of this age shimmered among humanoid, animal, and spirit forms; but they always had the same emotions and sensibilities as humans"[2]

Of course, oral tradition is also rich in shorter, lighter and sometimes amusing stories. They often play on people's foibles. They enrich childhood and adolescence by introducing the listeners to anger, joy, sarcasm or distance, both in time and space.

"We have no word for love in Lushootseed, so it is good to be able to recognize the signals," says Vi Hilbert.[3] Some stories transform into songs, sometimes rather burlesque ("There is no barnacle in my bathtub," for example).

Coyote is—or at least believes he is—an important character.

---

1. Vi Hilbert, *Haboo*, Univ. Washington Press, 1985.
2. Vi Hilbert, *Haboo*, Introduction p. XIX.
3. Vi Hilbert, op. cit., p. IX.

He will allow a suitor to marry his fleet-footed daughter only if he can outrun her. The first candidate doesn't even come close. The second is Raven, also eliminated, followed by Mink. One should be wary of those two; they are of the same ilk as Coyote: tricksters, full of mischief. Not to be taken as role models—they are immoral, deceitful, insolent, often victims of their own duplicity. Nevertheless, they accomplish extraordinary achievements; for example, they spread daylight over the world. They are true civilizing heroes.

Vi Hilbert points out that Mink and Raven are local, Lushootseed figures, while Coyote belongs to sagebrush country, the area east of the Columbia River. Trade and marriages between tribes account for Coyote's presence in the west.

This remark by Vi Hilbert opened my eyes to the idea that different creatures could embody very similar deities. It is well known that the tribes traded goods and ideas, as described by Carol McMillan's poem:

> From Mexico and Western shores the traders came
> With flutes and birds and tinkling bells—
> With gentle words.
> When Kokopelli played his songs
> The canyon filled.
> He brought along bright parrot feathers
> Stories, jokes, and messages
> From distant folks.[4]

Prehistoric petroglyphs of Kokopelli are found throughout the US Southwest. Kokopelli originates with the Zuni and Hopi Indians. He is hunchbacked, often shown with erect phallus, and plays the flute, like Arcadian shepherds and the god Pan. Some of his characteristics are reminiscent of Coyote, Raven, Mink and even Blue Jay, all standing for the divine trickster, the "most ancient of all mythological figures," says Paul Radin,[5] and assuredly one of the most relevant in today's world.

---

4. Carol McMillan, anthropologist, teacher at Omak Community College, 1990.
5. Paul Radin, *The Trickster: A Study in American Indian Mythology*, Shocken Books, New York, 1988.

Coyote finds in Fox a partner even more clever than himself. Fox sometimes comes to Coyote's rescue in dangerous circumstances. But he shows up only from time to time. However, one could not do without the other, at least according to this creation story from the California Achuwami Indians: the Creator appeared as a small cloud which condensed in the shape of Silver Fox. Then came a fog that turned into Coyote. Together Silver Fox and Coyote prepared the earth for humans, vanishing soon after their arrival. Even when condensed into a few words, the story retains its fascinating power.

One could now beg: "Haboo!" The reader might now want me to relate more stories from Vi Hilbert's repertoire. I shall do nothing of the kind, for I must go on with my own story.[6]

Let's just not forget that Coyote is the "opener," the inventor, the inverter. Anyway, he is not far from me: I have seen it many times over the past few days. Near the Okanogan river. And I know that behind me, on the right, behind the sagebrush, sasquatch was also seen a few months ago.

Finally, I was forgetting that Coyote also has the power to create, as does the devil today. Coyote is often the instigator, but also the pretext, the excuse. But we shouldn't worry about it. That's what Andy Joseph thinks, although he would not of course pretend to speak on behalf of his tribe. Incidentally, that's my advice too.

---

6. Sometimes, I listen to Vi Hilbert's recorded stories. Her soft and low voice reminds me of her slim silhouette, her dignified demeanor, her skilfull mastery of storytime. Her research, teaching and shamanic knowledge make her a truly remarkable spokesperson for her people.

Chapter 47

Through the Stone Gates

A few days ago, I was invited to a graduation ceremony. Wearing a borrowed gown and mortar board, I blended in with the faculty of Omak Community College. I sat next to an enormous drum set on the stage of the auditorium where the ceremony was held. Five Indians sitting crosslegged intoned a powerful song accompanied by the irresistible rhythm of the drum. My whole body vibrated to the point that my sense of balance was affected. Fortunately, the song was soon over. Thelma and Carol gave me a small drum decorated with the image of a brown bear. They knew that I was particularly fond of the bear, so close to the Indians, the sasquatch and all wild men.

It would have been the time to present the short speech that I had prepared for the occasion, just in case. But I found myself moved as well as slightly disoriented and, to my embarrassment, I could only utter a few banal comments. At least, I was brief.

After the ceremony, I entered into conversation with a Nez Perce student, accompanied by his father who, hearing that I was planning to travel to Nespelem, emphasized the sacred nature of the heart of the reservation. He suggested that behind the (Christian?) miracles, might be hidden other, more ancient phenomena.

The Fourth of July holiday was near, and the thought crossed my mind that the Indians might cloak their traditions in the framework of that, and perhaps other holidays.

On the road to Nespelem, images, advice and memories were jostling through my mind. But suddenly, I was past the rocky gates that might have slowed me down. Time flowed quite normally. I reached Nespelem without incident, even slightly early.

Adeline Fredin wears her long graying hair in a pony tail. She shares an office with a coworker and a secretary, all Native. She has a stern look, a commanding tone of voice, decisive gait and gestures. As a historian and archaeologist, her current responsibilities consist of examining land use and resource exploitation projects on the reservation, such as logging and new construction sites. She sees

Graduation ceremony at the Omak Campus of the Wenatchee Valley College (June 11, 1994). The author is the second from the left.

to it that sacred sites are protected; she also carries out archaeological digs.

Adeline greeted me with some mistrust, warning me against hasty interpretations of Native legends.

I told her that I had difficulty understanding some of Charlie's stories. I sometimes asked his neighbor, a Schuswap, for explanations. Adeline told me sharply that each tribe had its own culture and that one should avoid mixing them up. She then suggested some guiding principles: I should stick closely to the animals and the elements. She specified that the wind, moon, and the sun are to be my true ancestors.

I should go back as far as possible in time, all the way to the origins; she showed me reproductions of pictographs advising me to focus on "creation stories." As she said: "To fully understand the role which belongs to each animal, one must transcend their individual nature." I must thus ponder over the meaning of the spirit of the animal figures for today's Indians.

Adeline also mentioned that according to ancient tales, the Columbia used to flow in the opposite direction. Then, looking at me, she concluded that an approach to Indian culture through the animals seemed reasonable to her.

The interview was over. I had found some cautious support for my research.

Stepping back into my car, I thought about the tepees I had noticed on my right, driving into Nespelem. They were arranged in a wide circle. I stopped to take pictures, and as I did so, four youngsters sitting in the shade of an awning jokingly offered advice on the best angles. As I was about to leave, a young woman called me: "Hey, you! Want some Indian fried bread?" Bannock, or Indian fried bread, is a simple but a delicious treat. I was also offered a homemade, herb-flavored bean and ground-meat stew—far tastier than the usual, often pasty, chile con carne. Lila Whalawitsa was the cook, bright eyed behind her glasses and sporting salt-and-pepper braids.

I told her about the impact of the great drum at the Omak graduation ceremony and of my inability to remember my speech. She explained that the drum was the voice of the Spirit; that one does not prepare a speech: it has to come from the heart. How could one possibly communicate with strangers, with a wide variety of views and beliefs with a prepared text? She also told me about the forthcoming powwow and how it brought participants together at regular intervals, allowing them to release their emotions over a few days. Scattered families—some coming all the way from California, even Arizona—traded news, renewed acquaintances, and reminisced over past powwows.

To the side, in front of a tepee, two parallel tables were set up. Drummers and bell ringers sat at each end, opening up the camp. This was an important moment, not to be photographed. Each day of the week opened with a different song.

On Saturday, everyone would be invited to a big feast, salmon and venison to be on the menu. There would be dances, stick games with prizes of money, saddles, blankets, baskets; also storytellers, skilled in creating as well as preserving traditional lore. In the French countryside, there also used to be traditional stories, some of which were gathered in the *Roman de Renart* (twelfth century). Renard (Fox) reappeared more recently in Saint Exupéry's story of *The Little Prince* (1943)[1] Soon afterwards, I sent Lila Whalawitsa a copy of *The Little Prince* to read to her children.

---

1. The Little Prince is more than just a story for school children. Fox is a universal animal archetype; for example, Marcel Griaule and Germaine Dieterlen in their book *Le Renard Pâle* speak of the role of Fox among the Dogon of Africa.

When we met again at her home in Yakima, Washington, she showed us how while gathering berries the Indians celebrated the fullness of the season, communing with nature, joyfully living together one of the year's best moments. At the end of that encounter, Lila took my wife and me to a tepee so that we might enjoy a view of the universe framed in the opening at its apex.

In Gerald Hausman's words: "The tepee is both penis and vagina. Its form and meaning attest to it: yin and yang, cone and circle, mother as well as father, full of opening as well as thrusting energy."[2]

The tepee, the sweat lodge and the potlatch are mythological beings to the same extent as sasquatch or orca. They also possess an embracing power and a capacity to inspire and transform which, if it doesn't exactly put them on the par with the animals of the mythological pantheon, certainly makes them indispensable auxiliaries.

I was to return many times to creation stories and, as Adeline Fredin suggested, would pay particular attention to what they had to say. I soon noticed the continuing presence, particularly in Okanogan stories, of the creation still at work in today's world, a creation seen with respect and even reverence. Coyote often plays a leading role in this region, between the Cascades and the Rockies.

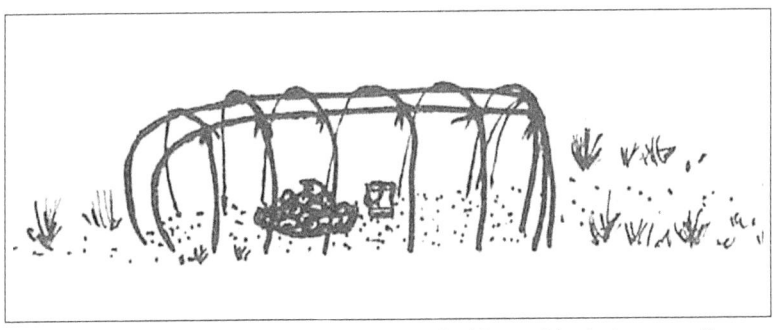

The sweat lodge: a birch frame covered with skins or blankets; a small oven to heat the stones; a container for the water which will be transformed into steam on contact with the hot stones.

---

2. G. Hausman, *Bestiaire des Indiens d'Amérique*, p. 60.

A variant on the Giving of Names story comes to mind. In it, the Great Chief (or Supreme God), faces the divine trickster with a dilemma. There are but two names left, he says, Coyote and Sweat Lodge. I choose Coyote, answered the trickster, and the Great Chief bestowed upon him the power to achieve extraordinary feats, beyond the scope of any of the tricks, good or nasty, that Coyote was so fond of. That's how Coyote came to free the Earth from the cannibal monsters.

Famous all the way from Alaska to Costa Rica, Coyote also figures of course among the Navajo. There, he convinces one of the giants who roam the world devouring children to come to a sweat lodge. In the dark and steamy confine of the lodge, Coyote announces that he is going to perform a miracle. He will break his leg and immediately fix it. Then with a stone, in the half-light, he breaks the leg of a deer and sings: "Heal quickly, O my leg!" The giant reaches and feels Coyote's leg: it is unbroken!

Coyote suggests to the giant that he should now break his own leg. The giant howls with pain. Coyote tells him to spit on it so as to heal it; he then escapes, leaving the child eater to his pain.

Ed Edmo, a storyteller linked to the Shoshone, Bannock, Nez Perce and Yakima tribes, grew up near Celilo Falls in the Columbia Gorge.[3] In 2001, a friend of mine heard Ed tell a tale similar to that of Coyote and the giant of the Navajos. Coyote now explained to

Coyote Stories

By Mourning Dove *(Hu-mis'-hu-ma)*

Edited and illustrated by Heister Dean Guie with notes by L. V. McWhorter *(Old Wolf)* and a foreword by Chief Standing Bear, Oglala Sioux

The CAXTON PRINTERS, Ltd.
CALDWELL, IDAHO
1934

Cover page of Coyote Stories, wherein Mourning Dove demonstrates her storytelling skills.

---

3. Edmo's links to these tribes are through either kinship or adoption or both.

bigfoot how to cross the Columbia River. All he had to do was to get rid of his legs for a while. He would then slide on his flat lower trunk to reach the other side. Here again, Coyote destroyed the Native giant's legs and left him to his agony.

Ed also told a number of anecdotes about bigfoot sightings. One day, his father had seen a footprint. On another occasion, some friends driving from Portland, Oregon, to Warm Spring, Washington, stopped to rest along the way. A bigfoot woke them up by shaking their car. In 1987, other friends saw a bigfoot looking through the windows of houses in the village of Wapato, near Yakima, Washington.

Edmo added that the Indians usually do not report such incidents as they fear that bigfoot might come to harm. A very common attitude, let me also add. Thus, as ordinary reality intersects the plane of mythological reality, we discover once more the enduring presence of the mythological realm relative to the everyday world.

Chapter 48

The Killer Whale

The American Indian bestiary of the Pacific Northwest gives a place of honor to the bear and the salmon. They are closely linked. The salmon is a manna from the sea for the bear and also of course for the Indians, both on the coast and up the Columbia. Men and bear enjoy salmon as a most nutritious food; when the stocks are low, both men and beasts suffer. Bears then raid people's larders, risking their lives in the process. When salmon are abundant, bears leave people alone, gorging themselves on dying fish during the spawning migration, and ridding the streams of rotting carcasses.[1]

Reports say that sasquatch also feeds on salmon. It has been seen standing or sitting in a river, catching fish as they swim by, just as bears do. Some say that sasquatch moves from one watershed to another to take advantage of the spawning migration, which varies by a few days from one river to another. All they have to do is to go from one tributary of the Columbia where the migration is over to another where it is still underway to reap a rich harvest.

The salmon and the bear (as well as the sasquatch) follow synchronous rhythms. As the light fades, late in the fall, bear and fish disappear, to be reborn a few months later.

Gerald Durrell, author and naturalist, points out that a bear does not really hibernate; it sleeps deeply, but its temperature stays around 95°F (35°C). It was long thought that its temperature fell to 41°F (5°C), but in reality, since its temperature is close to normal, a bear is quickly awakened when disturbed.[2]

---

1. Besides, spawning salmon have exhausted all their body fat in their upstream struggle, as well as much of their muscle mass, and their flesh is no longer appealing to human tastes.

2. Durrell comments that the scientist who slipped into a bear's den to take its temperature must have been rather daring.

Leaping out of its den, when disturbed or at the end of its sleep, a bear with its fur covered with moss and twigs looks like the very image of a sylvan deity.

Bear and salmon are examples of the cyclic character of the American Indian worldview. In that cyclical context, some steps are particularly important, namely the trials linked to the initiation to adulthood and their ensuing transformations. The rites of passage during which a child acquires a guardian spirit mark the beginning of a new cycle, that of adolescence. The teachings of the guardian spirit then become most valuable in preparation to future entry in the world of adults. We recall the importance of the role played by sasquatch, the guardian spirit of Andy Joseph (see chapter 21).

Before fully acknowledging the importance of sasquatch, I feel it is necessary to continue the description of the major figures of Native mythology. While bear and salmon are common to the Salish people of the coast as well as those of the interior, killer whale, or orca, belongs to the coastal tribes. Nevertheless, because of the geographical proximity and the spiritual kinship of the various people of the Pacific Northwest, it is familiar to all.

The first time I saw whales was in a summer afternoon in the Kenai Peninsula, in Alaska; they were belugas, swimming upstream against the ebbing tide in a small coastal river. There were two of them, one on each side, swimming close to shore. What an unforgettable sight! A quarter century later, I had a good look at belugas and sperm whales in the Saint Lawrence estuary, in Quebec, Canada. How could one ever be sated of a spectacle that brings such great pleasure and exciting satisfaction! Each time the occasion arises, I seek to renew this wonderful experience. Folk stories help to renew the links with these fabulous animals.

A Haida story relates how a suspicious husband discovered that he had married an orca woman. Every day, his wife went to the sea shore and beat her mat. A killer whale would appear and make love to her. After witnessing this scene, the husband put on his wife's clothes and, armed with a knife, went to the site of the tryst. The orca, penis erect, leaped from the sea. The husband cut off his penis as the orca swan away screaming. That night, the husband cooked the orca's penis, telling his wife he was preparing something very sweet. She ate it with delight. That's when the husband told her, "Your orca lover is very sweet isn't he?" She ran away and dived

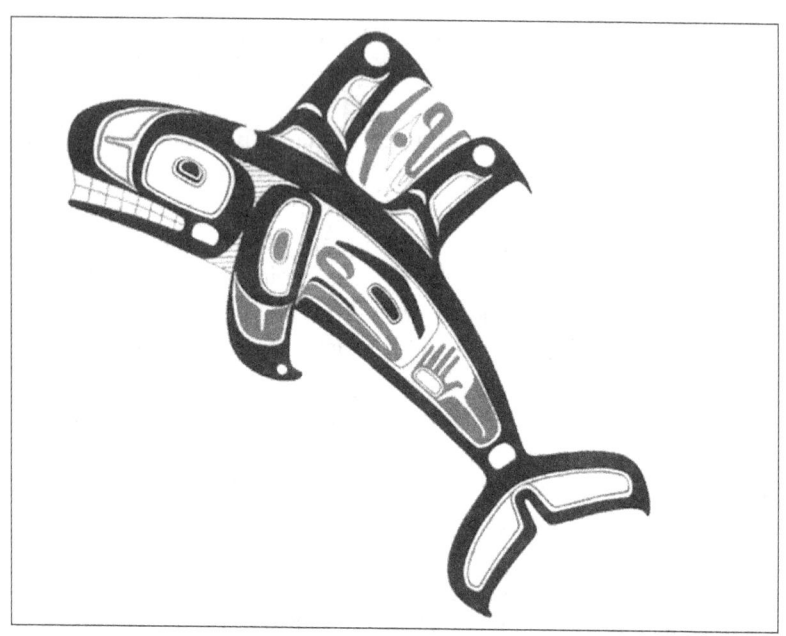

The killer whale, one of the major figures of the Pacific Northwest pantheon.

into the ocean. The husband then knew that he had married an orca woman born into human flesh. He saw her body transform into a reef when touching water. That body, the reef, is now called The Woman.[3]

Gérald Hausman notes that animals are "permeated" by humans through dreams, visions, drugs, the ecstasy of trance and dance, even art, where human hands give life to the animal spirit. However, he adds, to eat an animal, to become the same flesh, that is to know its essence.

Everything is interpenetrated. By dressing up as his spouse, for example, the husband identifies with her.

But the most beautiful and mysterious feature of the story is the identification of the young Haida woman to an orca woman. She looks like a human being and she behaves like one, but her heart is part of another race. As to the end of the story, it is of striking sim-

---

3. Gérald Hausman, loc.cit. p.58.

plicity. The transcendentally powerful woman returns to the water, to the mother, to its origins. Far from any pretending. She communes with nature; she IS nature. And if a human being sets foot on the reef called Woman, the Haida Indians are convinced that it wobbles, for supernatural beings do not wish human beings trodding upon their back.[4]

Among the coastal Indians, there is a plethora of stories about the killer whale. What sticks to my memory is how Haida sculptor Goolaslacoon (aka Ralph Bennett) represents the killer whale. In the Haida language, *Goolaslacoon* means "abalone fingers," which may also mean "mother-of-pearl fingers." The nacre from inside the abalone shell is commonly used in Pacific Northwest Indian masks.

One day a Native woman from Sitka, Alaska, told me: "One must grasp the structure of Indian representations; it is necessary, for example to understand the meaning of the whale in an Indian cemetery."

I did my very best, remembering the guiding role played by the "great fish" among many people on Earth. An agent of transformation and resurrection—like Jonah—or also as a scout on the path to immortality: those are the roles attributed to the killer whale.

It is while in Goolaslacoon's workshop that I glimpsed some of the meaning of the whale. His workshop stands out from other Haida wooden houses by its glass-fronted façade, under a gently sloping roof. Ralph heats his house with yellow and red cedar wood chips; the house is cosy and smells good. At work, Ralph starts with a song in his native tongue, whirls through a dance step, picks up his adze and carries on with the carving of a large halibut hook (halibuts reach up to five feet in length!). The elegant lines of the hook are embellished with a carving of a mosquito, all in honor of the prospective victim, the halibut.

From the inside, the workshop looks like the overturned shell of a ship, an inverted arch (Latin: *arca*), keel facing the sky, or also perhaps the skeleton of a killer whale (orca). Ralph tells stories from his family, especially from his father who, like him, was a sculptor of masks and totems. One of them comes to mind. It is about a young Indian who, with a friend, flouts the laws of the tribe. As a

---

4. Gérald Hausman, loc.cit. p. 60–61.

punishment, they are thrown into the ocean, where they fall for a long time, surrounded by orcas gracefully circling around them.

A perfect beauty robed in velvet, with a white spot above its eye; a living black-and-white symphony, the orca, its mouth studded with great ivory teeth, is as supple as a snake, its motion as fluid as that of a drop of oil.[5]

At first, the terrified guilty youths see their end approaching. However, the orcas invite them to join in their dance. At first surprised, then wonderstruck, they gradually improve their strokes and with time become increasingly graceful. They discover the lofty buildings which are the houses and the submarine cities of the orcas. They also discover the exactness and soundness of the laws in effect among those builders. They become more and more like orcas.

One day, they return to the surface and visit their old village, listening in on people's conversations. The orca youths are surprised to hear how much strife there is. They crawl onto the beach and enter the village. However no one either sees or hears them. They return to the depths, overcome their disappointment, and follow the teachings of the orcas.

The Bella Bella Indians, neighbors of the Haidas, speak of Qanelekak, who lived in the days when the Earth was nothing but water and ice. Qanelekak's torso was that of a human, but the rest of his body was that of a whale. One day, he ran his hands down his body and became fully human. As the ancestor of the Killer Whale Clan, he created the dog, which he presented as a gift to all people, to whom he attributed names, as he also did to the animals.

> The ovoid shape of the whale has been compared to the intersection of two circles, symbolizing the union of the world above to that below, the sky to the land.[6]

What I remember from all that is that there are two worlds mirroring each other. Which one inspires the other? It would appear that

---

5. Pierre Clostermann, *Une Vie pas comme les autres*, p. 230.
6. J. Chevalier and A. Gheerbrant, *Dictionnaire des Symboles*, Vol. 1, p 168.

the invisible universe of the orca would inspire, refine and even fashion the world at the surface.

The dual nature of mythological characters is an enduring feature of the American Indian universe. Speaking of Coyote, Gary Snyder remarked that it knew no clearly defined boundary between good and evil. Coyote is often attracted to the dark side, usually through some burlesque prank. Being serious is not its forte!

As to the killer whale, also endowed with a dual nature, its role is truly a serious one: it is a teacher.

A carved orca, sculpted by Ralph, adorns the totem stick of his son, a young computer science student. The orca is his son's guardian spirit. Through his father's work, the young man learns under the guidance of a demanding master, the orca. It is through stories, songs, working on masks and carvings, that the young man absorbs his father's lore. He and his father follow their respective creative paths.

Chapter 49

Bigfoot is Everywhere

The Coeur d'Alene tribe, in Idaho, told how Coyote, in addition to contributing to their culture by introducing the salmon, creating fishing sites and teaching many skills, had also left alone, without trying to transform them, dwarves and giants who continued to live in the forests and the mountains. Giants were to be found in all North American tribal traditions.[1]

In the Coeur d'Alene legends, the giants exhibited an aggressive behavior. One giant, for example, leaned against a wooden house to shake it, without harming its occupants. It often seemed that the giant took pleasure in scaring people. This type of behavior, sometimes accompanied by rock throwing, occurred until recently. It is now attributed to bigfoot, who has merged with the traditional giant. John Green speaks of loggers, besieged in their cabin, who had the fright of their life, although their attackers left without hurting them.

After 1960, there are fewer reports of such aggressive incidents. Bigfoot is now described as an essentially peaceful creature, frightening by its size, but discreet and most likely to slip quietly into the forest.

The ethnologist James Teit reported that the Spokane and the Coeur d'Alene claimed that the evil-smelling giants, dressed in buffalo skins, had the capacity of transforming into trees or bushes. Some giants dressed in bear skins, lived in caves and were fond of eating fish. Sometimes they would abduct women from neighboring tribes. The Okanogan Indians said that the giants stank like 'burning gunpowder.'

---

1. There seem to be as many giants in South America. The sacred book of the Maya-Quiche, the Popol Vuh, speaks of the important role played by the three giant inheritors of the solar cult. They stand in opposition to the Gemini. The Maya civilization florished from AD 200 to 1220.

The Flathead, a Salish people from Montana, had an attitude of noninterference with the giants; they ignored each other as much as possible. The giants were hunters, like the Indians. That they were better at it was not due to their being better equipped, but to their greater strength and cunning.

People of the Pacific Northwest had a variety of names for the giants. They were known as Tree Strikers, Whistlers, Stick Indians, Cockleshell Men; it was not uncommon to find one gathering clams and mussels. These giants avoided whites, whose odor they disliked. The Pacific Northwest giants appear under a wide range of variations. The anthropologist Wayne Suttles attempted to compile a list of all the creatures covered under the concept sasquatch/bigfoot. He found that the concept applied to widely different beings.

In the past, giants were common to all civilizations. The British Columbia Kootenay Indians called them Eka; for the Lakotas, they were the Iya; among the Inuit of Hudson's Bay, the Tuurnngaq set traps fatal to hunters; among the Seminoles of Florida, Tall Man broke

A Kwakiutl heraldic pole that shows Dzonokwa (the cannibal woman). Dzonokwa (also D'sonoqua) is the main crest of the Nimpkish Indians.

tree branches to make masks. The Okanogan called them Twsainaitem and said they used flutes imitating the songs of birds, the calls of squirrels. In most cases, giants are characterized by their harmful behavior. They act like the ogre of European fairy tales, carrying little children deep into the forest to devour them. They are called upon to threaten disobedient and unruly children. The giantess Dzonokwa of the Northwest Coast Kwakiutl is particularly frightening, because of her size and ugliness. She is the cousin of Windigo, the giant cannibal of the Ojibway of Ontario and North Dakota. The giants, male and female, are symbols of brute force, one of the many facets of their character. They must step aside for civilization and it is Coyote's task, as the civilizing hero, to get rid of them.[2]

The cannibalistic giants have special characteristics. For example, it was thought that the Windigo was a human who, having tasted human flesh, mutated into a monster who roamed the forest, ready to devour lost hunters. However, a human being could also fall under the influence of the "Windigo spirit" and engage in cannibalism. Whoever was so possessed was often executed to protect the tribe.

Among the Kwakiutl the impressive Hamatsa rituals (the cannibal dancers) are performed during the ceremonial season, in the winter, by the Hamatsa secret society. The new initiates, their spirits taken over by the cannibals, move restlessly and bite other dancers. They are then taken to the forest for a few days and brought back to the longhouse at night, with the whole village in attendance. The initiate is introduced to the house through a whole in the roof; his head and wrists are decorated with hemlock branches. He whistles as a mark of savagery; he sings and shouts, asking for human flesh. The spectators try to catch him, in vain. The Hamatsa is caught only the next day. His dance becomes quieter; his bouts of savagery rarer; his companions calm him down, dancing slowly, wearing bird masks. Finally tamed and freed from the evil spirits, the novice, wearing a blanket and an apron, dances quietly with the spectators and other participants.

The cycle civilization/wilderness/civilization allows people to appreciate the power of wilderness and of the Hamatsa order which, thanks to carefully planned rituals, succeeds in transforming the initiates.

By using the term sasquatch/bigfoot for a variety of different beings, from the bogey man to the cannibal giant, one risks masking the specific nature of different categories of giants or wild men. What can be clearly recognized as an essential trait of the wild man is the transformative power it exerts over those who meet him.

For the Athapaskan tribes, sasquatch is said to appear during times of difficulty so as to enable Native communities to regain their harmony with Mother Earth.

---

2. The woman-kidnapping giants are common in North America. This is a category well known to all, as it exists everywhere.

I will return below to the importance of sasquatch encounters for the inhabitants of North America, particularly the Indians. I will also offer some hypotheses to explain the meaning of the sightings proliferating over the whole of the United States.

For now, let's leave the Pacific Northwest behind. We have already ventured beyond the borders of that area, into Montana as well as into North and South Dakota. We were then only following the Idaho and Washington tribes hunting the buffalo among their eastern neighbors. We have also crossed the Canadian border, a nonexistent boundary for the Indians, to mention some practices of the Ojibway.

This time, we will reach out to the Canadian Northwest Territories to visit the Chippewa tribes. Some of them are nearly sedentary, but hunt during the spring and summer, camping in the wilderness. This information came from the works of socio-anthropologist Henry S. Sharp, married to a Chippewa woman. For many years around 1970 Sharp accompanied caribou hunters gathering meat for their families during the winter. A few words will suffice to introduce the reader to the general atmosphere that prevails during the hunting period.

First of all, we note that power does not attract any of the members of the tribe, contrary to what takes place among whites, especially in the realms of politics and economics. Competition exists, but it stops where the interests and goals of neighbors are threatened. This attitude reflects the egalitarian nature of Indian societies.

Of course, the Chippewa make no clear distinction between the natural and the supernatural worlds. The word *inkoze* covers the basic notion of "power," the strong "medicine" which allows prophecy, the curing of the sick or success in the hunt. The Iroquois called it Orenda, with an evil counterpart called Otgon.

Among the Chippewa, one should not speak of inkoze.[3] "It is knowledge about the operation of the universe revealed to humans by supernatural creatures in dreams."[4]

Anyone claiming to possess that power would be considered an impostor.

---

3. For some individuals among American Indians (particularly the Nez Percé), it is bad form to speak of bigfoot.
4. Henry S. Sharp, *The Transformation of Bigfoot*, p. 37.

In the end, one should remember that American Indians often speak in a devious manner, enigmatically, through their myths. The Chippewa seem to have few stories to tell. Henry S. Sharp relates an incident that happened during a hunt for caribou, a widespread and common animal and a major figure of local mythology.

Quite by accident, a young Chippewa had killed a wolf cub roaming in the vicinity, still too young to be a menace to anyone. Henry Sharp was unhappy with this and refused the skin, which was offered to him, because there had been no good reason to kill the animal. An Indian took Sharp aside and told him the story of a herd of caribou kept by Raven (another divine prankster) in a corral. Raven, mistrustful of one and all, kept a loose watch over his herd, day and night. Nevertheless, the swiftest of all animals, White Fox, managed to make a hole in the enclosure, through which the caribou rushed out.

A drawing by Wayne Moore (Elma, Washington, 1994).

In my view, the caribou—meaning the people—can shut themselves in, following laws or principles set forth by the supreme chief, Raven. But there is another aspect to the chief, that of White Fox, which it pays to be aware of.

As for Henry Sharp, he remained confused, but concluded: "Sometimes, even when facing the strangeness of the other in fieldwork, context and shared experience reveal more than words and logic."[5]

Following these few words, which helped to define the context of mythological reality, it is now time to deal with bigfoot/sasquatch. It turns out that the Chippewa originally had no knowledge of the

---

5. Henry S. Sharp, op. cit. p.74.

hairy giant. However, in June 1975 a well-to-do American tourist came to fish and hired a young Chippewa guide. Corky, the tourist, told everyone stories about bigfoot. He and his wife were fearful of the grizzlies whose tracks had been found nearby in the spring. The couple seemed to hide its fright of the bears by spreading stories about sasquatch. Their guide translated the term *bekaycho*: *be* = he/she, *kay* = foot, *cho* = big.

Although until 1975 the Chippewa were unaware of the existence of bigfoot, they were well acquainted with the bogeymen, who besides their role in frightening children, had a diabolical influence on adults. We note, however, that whenever asked, the Chippewa would automatically answer that they had never seen one. Who were these bogeymen?

There were many kinds, but they all shared the same characteristics: they lived alone in the bush, they prowled around the Indians' houses, had an evil temper, and were prone to stealing and kidnapping. But they were not at all like the sasquatch in terms of their size or smell.

In 1984, a Chippewa village saw with astonishment and mistrust the activities of a paramilitary group that set up a training camp nearby. Dressed entirely in black, carrying survival gear and armed with long knives, they wouldn't respond to the villagers' greetings. The latter began to think of them as bogeymen and on occasion even shot at them.

For the Chippewa, any kind of bogeyman was dangerous. Bogeymen also filled the role of "scapegoat," and were held responsible for all kinds of reprehensible behavior: conjugal violence, abuse of children, etc.

Over the years following the departure of Corky the amateur fisherman, Bekaycho became more and more popular. The Chippewa guide had transformed it into a white prowler loitering aimlessly around camp sites; it was a menace for women and children, stole meat, robbed caches and spoiled supplies of caribou meat.

It is quite likely that in an effort to exorcise his fear—undoubtedly well founded since he and his guide had to face a grizzly during a hunting trip—Corky took pleasure in telling bigfoot stories. He was also probably trying to comfort his wife, who found herself uneasy living in the wilderness. Of course, for the Chippewa the term wilderness had no meaning since it was the environment in which they had lived for untold generations.

The Chippewa had found a new use for bigfoot, who as Bekaycho had come to include the notion of a bogeyman. This recent transformation goes completely against what one might have come to expect. It follows a path exactly opposite to the transformation of bigfoot by white Americans, who took a number of giants, bogeymen and cannibalistic wild men to end up with sasquatch/bigfoot, an enigmatic and silent being, endowed with prodigious but peaceful strength, infinitely adaptable, comfortable in the wilderness, as well as near towns and farms—a being who knows how to keep out of the way of men and their vehicles. There is no doubt as to its natural instinct and practical intelligence. Further, it is always fully under control of its behavior and even capable of controlling the perceptions of those who see it—hunters rarely shoot at bigfoot.

This new bigfoot/sasquatch, a blend of various beings arising from Indian traditions, has taken over the psyche of both whites and Indians.

It would be very tempting to simplify: the wild man bigfoot tradition of a parallel race of relict giants, as brought to life by the Patterson film, joins the American Indian tradition of mythological giants into a single entity. However, I think that to blend the two streams would be misleading. The American Indian sasquatch has its own specific properties, including a spiritual dimension.

Chapter 50
Dzonokwa and the Potlatch

When studying the masks of the Pacific Northwest tribes and trying to decipher their complex symbolism, the great ethnologist Claude Lévi-Strauss declared, as a working hypothesis, that the features of the masks that he thought most noticeable had no intrinsic significance, or at least that their inherent significance was by itself incomplete. Every effort of interpretation would thus be in vain.[1]

Pacific Northwest traditions feature a giantess who, of course, attracted Lévi-Strauss' interest. She is the subject of impressive representations through masks and totems of striking appearance and dimension.

Claude Lévi-Strauss remarks that the Fraser River (British Columbia) tribes call her Sasquatch, or Tsanaq, a black giantess, with heavy eyebrows, deeply sunken eyes, long and thick hair, big lips and frowning mouth, the hollow cheeks of a corpse.

The tribes of the British Columbia coast and their neighbors know her under the name of Dzonokwa, usually female; there is also a Dzonokwis, a marine equivalent of the forest giantess.

The Makah, at the extreme tip of Washington State's Olympic Peninsula, tell the story of Salmonberry, a little girl playing on the beach with two friends while her mother was gathering clams. Salmonberry promised to return home before sunset. But she forgot. The giantess, Basket Woman, picked up the children and took them to her lair to roast them.

The worried mother returned to the beach and found, in the sand, the footprint of the giantess. Her tears fall on a clam, which gave birth to a handsome young man, Clamshell Boy, who set out to find the children. He found Basket Woman, flattered her and talked

---

1. See Claude Lévi-Strauss, 1982. *The Way of the Masks,* trans., Sylvia Modelski, 1982, Univ. of Washington Press, Seattle.

her into admiring herself in the water; leaning forward, she fell into the sea and drowned. Clamshell Boy untied Salmonberry and her friends. He helped them peel off from their eyelids the pine resin with which Basket Woman had sealed her prisoners' eyes.

There was great rejoicing in the village-at-the-end-of-the-land. The People held a celebration of potlatch where they feasted and danced and sang. Everyone exchanged gifts, and told tribal stories. It is in this way that the Makah still observe their important occasions. It was Clamshell Boy who brought potlatch to the People of the Cape [Flattery].[2]

Thus ends the story.

Today, in the summer, the Makah paddle their canoes towards the east-southeast, from port to port. They gradually reach the waters of the Quinault fishing tribe in Puget Sound. This is how the young Natives maintain the navigating skills of their ancestors, readying themselves for the challenges of the whale hunt. Since 1997 the law has allowed them to kill one whale each year. In this way, the government has made it possible for them to renew their links to a cherished tradition.

The Dzonokwas, antisocial spirits with weak eyesight, attempt to blind children before devouring them, thus jeopardizing the future of the tribe. As a story for children, it helps to teach them obedience to family rules, respect for the elders, and fear of the unknown—the forest, the night and all those dangers that threaten children.

There are also long and complex stories that describe more paradoxical activities of the giantess. She no longer appears merely as the witch who the children must get rid of in order to survive; she transcends her role as monster of the darkness. She is said to get hold of the children as if they should belong to the group of the wife rather than that of the husband, as suggested by Claude Lévi-Strauss. The matrilineal organization of society—among the Kwakiutl, for example—surfaces through these tales. Though a kidnap-

---

2. Terri Cohlene, *Clamshell Boy*, p. 30.

per, Dzonokwa also supplies potlatches with "coppers." To the great surprise of European newcomers, these crafted copper slabs were considered by Pacific Northwest tribes to be their most valuable possessions.

To grasp the scope of Dzonokwa's impact, it is essential to recall the role of the potlatch in Pacific Northwest Native culture.

The potlatch, held on special occasions, such as the installment of a new chief, or the "feast of the dead," consists in gathering a profusion of gifts. During a whole year, perhaps two, the family and the whole village of the new chief amass gifts for their guests from the next village, whom they will entertain for three to 15 days. The prestige of the hosts and their new chief is enhanced by their lavishness, at the expense of their resources, diminished to the point of penury. There is little risk though: the guests, to keep up, now prepare to be as generous in welcoming their former hosts. The potlatch transforms potentially aggressive instincts into friendly competitions, a form of trade or barter. A barter, indeed, seen as a menace to North American mercantilism: in 1884, the Canadian government banned potlatches as a means of suppressing Native culture (a ban revoked in 1951). As a consequence, large meetings were discouraged, preventing the rise in prestige of individuals or families whose status as "noble houses" would have been enhanced.

It turns out that the Pacific Northwest Indians were not the only people practicing the potlatch. In 1902 Maurice Leenhardt, a young protestant missionary in New Caledonia, described a similar ceremony, the pilou, among the Kanaks. Here is how he describes it:

> the summit of Kanak society is not a hierarchical head, a chief, it is the pilou itself: the moment when allied clans join together, with dances, speeches, the exaltation of ancestors, totems and the spirits which are the source of life, its powerful supports, the very basis of society.
>
> Should pilous be discontinued, their society will lose its cohesion and disintegrate.[3]

Dzonokwa, provider of coppers and symbolically of light and

---

3. Maurice Leenhardt, *Le Pilou, moment culminant de la société*, p. 67.

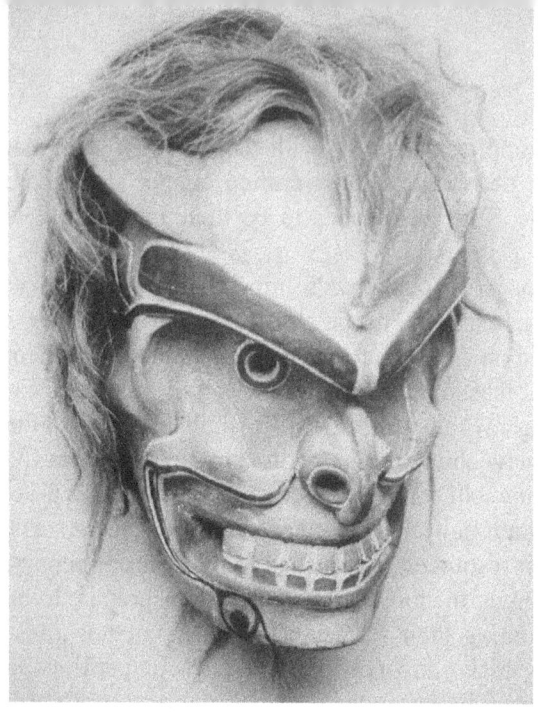

Dzonokwa, the "wild woman of the woods" (Kwakiutl dance mask).

knowledge, seems linked to the granting of a crest—the equivalent of heraldic arms—or the affiliation to a secret society, as among the Kwakiutl.[4]

The carver Beau Dick, author of a Dzonokwa mask (1993) specifies that the mask of the giantess may also be held in the hand of a chief during a speech. The mask then means that he has succeeded in acquiring the knowledge and powers granted to Dzonokwa by the forest. In the end, says Beau Dick, "Dzonokwa is the most important crest owned by a chief."[5]

At this point, it seems appropriate to quote the views of the famous ethnologist Franz Boas on non-European cultures:

> Anyone who has lived with primitive tribes, who has shared their joys and sorrows, their privations and their luxuries, who sees them not solely as subjects of study to be exam-

---

4. The study of the North American Native secret societies is well beyond the scope of the present study. According to Boas, it seems that the crests of the clans and the insignia of the secret societies were obtained in the same way.

5. Gary Wyatt, *Spirit Faces*, p. 74.

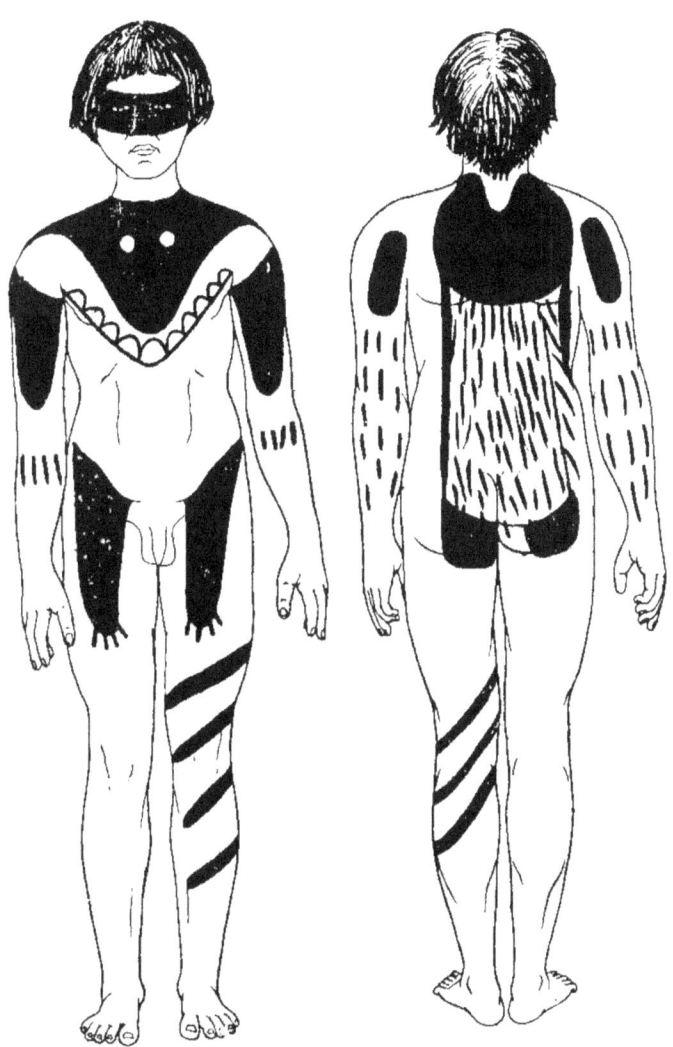

Body painting representing the bear (Kwakiutl) from Franz Boas, *Primitive Art*, p. 250.

ined like a cell under the microscope, but feeling and thinking human beings, will agree that there is no such thing as "primitive mind", a "magical" or "prelogical" way of thinking, but that each individual in "primitive" society is a man, a woman, a child of the same kind, of the same way of thinking, feeling and acting as man, woman, or child in our own society.[6]

With this attitude, Franz Boas successfully approached and penetrated the cultural universe of the people of the Pacific Northwest. Without understating the difficulties he encountered, he managed, thanks to his patience and rigor, to share with his readers the rich and unfamiliar traditions of the Salish Indians, among others.

On the stage of the great mythological theater, the actors—giants, bears, sasquatch, Indian, shaman—overlap and become interchangeable.

Sasquatch/bigfoot sometimes refers to a physical entity, but other times seems far away from any zoological equivalent. Nevertheless, the various figures illustrate the transformation resulting from a new birth (in the manner of Dionysos—the twice born) The sleeping bear comes out of winter as it exits from the mother cave. The shaman becomes, after his retreat and trials, his guardian spirit. All become Others.

The salmon, through its life history, summarizes the heroic phases of the cycle of transformations.

The killer whale shows the way to build for oneself and for the community. It is a signpost on the path of creation.[7]

Brotherly Coyote (often a false brother) explodes in coarse laughter. He becomes less and less inspiring and reliable. However, by jostling and goading he prepares a new way, which sasquatch heralds or merely suggests.

---

6. Franz Boas, A Wealth of Thought, p. 33.
7. Incidentally, thunderbird, the epitome, with the eagle, of a solar being, fishes it out of the sea and feeds on its flesh!

Chapter 51
The Giants

The works of ethnologist Franz Boas have inspired many others. A white investigator, professor and sculptor, Bill Holm, examined in exacting detail the stylistics of traditional carvings.[1]

Holm carved a nine-foot-tall (2.80 m)Dzonokwa sculpture, holding in each hand a kind of irregular shield decorated with black and white stripes. The two shields appear designed to fit into each other, like parts of a jigsaw puzzle: perhaps a symbol of how the union of opposites might depend on the good will of a giantess torn by conflicting tendencies. At least that was the impression Holm's sculpture left me with.

Actually, this reproduction of Dzonokwa was inspired by a "ridicule pole," dating from around 1912, found on Gilford Island, British Columbia. The right to represent this being through masks and totems or on ceremonial cloaks was the hereditary privilege of some noble Kwakiutl families.

Sketch of a Kwakiutl mask of the wild man of the woods. (Rob Butler, 1999; by permission of the artist).

When the pole was raised, Dzonokwa's gaze was fixed towards the house of the owner's in-laws, to shame them into remembering a wedding debt. In 1915, after the debt had been redeemed, the totem was rotated so that the giantess's figure would face the ocean. Coppers where then placed in Dzonokwa's hands to show that the debt had indeed been paid off. The giantess had successfully played her role as an instrument of reconciliation.

---

1. Holm's book *Northwest Coast Indian Art* (1965) has become a classic reference.

Male giants also lived with their sisters in the forest. The Iroquois spoke of Dehotgohsgayeh, the "two-faced" giant who lived in darkness. One half of his body was red, the other black. He wore the skin of a gigantic bear as a cloak and wielded a club made of the trunk of a walnut tree. His role was to protect humans against evil. We notice how the dual nature of the giants reappears again and again. Could the giant be the representation of some primordial deity?[2]

The sculptor Ralph Bennett explained to me that the bear was the model for the shaman. One might even venture to say that it actually gave rise to shamanism. In the old days, said Ralph, an Indian had followed a sick bear into the woods; he noticed that the bear was scratching the bark from this and that tree and eating it. As time passed, the bear regained its health and its strength. By following the animal, the Indian learned about the curative properties of various plants and barks, which he stored in his medicine bag.

Dehotgohsgayeh, the Iroquois giant, is sometimes called "Wry Face" because he is insolent and fond of irony. This attitude may sometimes appear paradoxical, but actually it is a way of helping people through a form of beneficial shock treatment consisting of salutary awakenings and stimulating responses.

I cannot resist quoting here from a popular dictionary (the Larousse of 1948), which defines a colloquial French use of the word Iroquois: "A person with bizarre habits: what an Iroquois!" Of course to a European, many Indian tribes would indeed have "bizarre habits;" there are "medicine societies" under the patronage of the bear all over North America.[3] Similarly, the club, a highly symbolic object, appears in a great many traditions. Among the

---

2. After all, behind Rabelais' Gargantua, there are glimpses of other, more ancient giants. The people of the Argonne area of France spoke of a legendary giant called Tord-Chênes (Oak-Twister). The geographers of antiquity, Mela, Ptolemy, Strabo wrote about a Mount Gargan in the French Alps, "Garganus Mons." Was there some connection between that hill and Hercules, also armed with a club?

3. The Bear Dance of North American Indians has its counterpart in Europe as a ritual celebrating the end of winter. During the cold season, both bear and man hid underground or in caves. The wild man's calendar was in step with that of the bear.

Celts, the club has the dual power of killing one's enemies as well as resuscitating one's friends.

In the South, among the Seminoles of Florida and Oklahoma, Tall Man, a gray-maned giant smelling like a rotten swamp, was said to break branches off trees to make them into clubs.

In Louisiana and Florida, there have been frequent reports of wild beings, similar to bigfoot. Covered with long reddish hair and stinking to high heaven, they are known as skunk apes.[4]

The story of Gispawaweda, from the Tsimshian tribes—neighbors of the Haida and the Bella Coola of British Columbia—emphasizes the role played by the bear as a master and teacher. Gispawaweda learned from a black bear how to make a canoe and to fish for salmon. Returning to his village, he now looked and acted like a bear, having lost the power of speech and refusing to eat cooked meat. After being daubed with ointments, he recovered his human shape as well as the habits of a civilized being. However, each winter, while his fellow villagers went without, he would join his friend the bear, who would feed him salmon.

This tale brings out, on the one hand, the merits of solitude, animal nature, and the need for a personalized initiation; on the other, the need for sharing, communication and civilization, which result from living in the society of men.

Ralph Bennett equates the term "sasquatch" (originally *saskehavis*—a Salish word) to *Gaagit/Gagid/Gagixit,* terms used by the Haida. He estimates that the creative power of Gagid is as strong as his gift for destruction. Bennett interpreted this duality in a mask that he carved about 15 years ago, using contrasting colors, red and black, and crowning the human visage with a fabulous bird of great power, the Thunderbird.

Haida artist Reg Davidson carved a mask of Gagid in 1988 and "danced" it many times. The Gagid dance is performed in two stages. In the first part, the dancer moves wildly, scaring the audience. Then caught with a cedar bark rope and tamed, Gagid recovers his human nature.

An ancient memorial pole from the end of nineteenth century,

---

4. For more information about the skunk ape, see Michael Newton, *Florida's Unexpected Wildlife,* Univ. Press of Florida, 2007.

erected in the graveyard at Alert Bay (British Columbia) provides a spectacular, 16-foot-tall (5.0 m) example of the connection between Dzonokwa, the Wild Woman, and the Thunderbird. It shows the head of Dzonokwa surmounted by the Thunderbird with wings spread out.[5]

Bill Reid, one of the most talented Haida sculptors, created a monumental piece, Raven and the First Men, part of the collections of the Museum of Anthropology of the University of British Columbia, in Vancouver. The sculpture is based on a Haida creation story which features Raven, a famous trickster which Claude Lévi-Strauss calls The Deceiver.[6]

The story starts with Raven having just landed on a long sandspit. The waters, which had long covered the whole world, had recently withdrawn and Raven had stolen the light from an old man who had kept it hidden in a box.

For once, Raven was not starving. However, he had other appetites to satisfy: his curiosity, his covetousness, and his insatiable thirst for meddling and playing tricks on the world and its creatures. Standing in the sand, he was soon bored. Then, out of the corner of his eye, he saw the white shell of a giant clam sticking out of the sand. Hopping over, he saw that the shell was crawling with small creatures, cowering in terror under his gigantic shadow.

Raven spoke to them in its most melodious and irresistible voice; curious, the small creatures ventured out of the shell. With long dark hair and flat features, they were the original Haida. They quickly learned from Raven who, however, rather short on patience, soon tired of teaching them. Besides, there were only males among them. Placing chitons[7] between the legs of some of them, he created some females. Raven's great game was now underway; it goes on to this day.

From then on, the children of these small creatures would prosper; they would construct and create, fight and destroy, their rich

---

5. This magnificent sculpture may now be seen at the Royal British Columbia Museum in Victoria, British Columbia.
6. Reid, Bill and R. Bringhurst, 1996, *The Raven Steals the Light*.
7. Chitons are mollusks commonly found in the intertidal zone, They bear a series of protective plates on their back and their lower fleshy part look like a woman's genitals.

and complicated existence following the rhythm of seasons and the tradition of rituals.

However, today the storms have overthrown the heraldic poles, the villages have disappeared, the people have been decimated by disease and only a few still dwell in the land. "Hasn't the time come for Raven to look for another clam?" asks Reid.

Every time I recall in my mind's eye the imposing sculpture of the Raven and the First Men, carved in yellow cedar, I also hear Bill Reid's question, arising implicitly from his masterpiece. The story, taking place in Haida Gwaii, (officially called the Queen Charlotte Islands[8]), filled my mind like an obsession. It was to have an unexpectedly comforting sequel.

Over the last weekend of May 1994, I found myself in Harrison Hot Springs, British Columbia, as a participant in a forum devoted to bigfoot, whimsically named Bigfoot Daze. This yearly meeting was held in a hotel on the shores of Harrison Lake and brought together a variety of speakers for two days of presentations and discussions.

The crowd is intimate and friendly; John Green is a local resident and his reputation as a pioneer bigfoot researcher draws a wide participation. Other veterans join him: René Dahinden and Grover Krantz, for example. However, there are no American Indians participants, either as speakers or spectators.

The presentations are dense with information, but the wide range of topics covered and the concentration required to grasp the different regional accents leave me, at the end of the day, rather exhausted. It's raining over Harrison Hot Springs. The lake is a curtain of gray mist. My friend Fred Bradshaw and I walk back to our B&B. As the rain intensifies, we take shelter under a wide portico. An Indian, probably in his fifties, has already found a refuge there from the rain. I saw him on the street yesterday; Fred told me that he was a wise man and a healer, a shaman in other words. He has a good reputation.

---

8. Although officially gazetted as The Queen Charlotte Islands, the original appellation Haida Gwaii is gaining ever wider acceptance, and there is talk on the islands of a petition to the Canadian Geographical Board for an official return to the Haida name.

The rain-imposed pause is a welcome rest. We begin chatting and I describe the nature of my research and my interest in sasquatch. I mention Bill Reid's sculpture and my obsession with the question it raises. After some reflection, the Indian answers:

> Raven is perched on the clam...Above him one should imagine another being, invisible, but nevertheless quite present...another being...he is called Sasquatch.

I am flabbergasted! "Sasquatch? Above?"

Yes, says the shaman as he leaves and wishes me good luck. We return to our lodgings and, at my request, Fred confirms that I heard the words of the medicine man correctly. There is no doubt: the wild man, the Master of the Forest, is biding his time.

Only a few days later, I received a letter from a shaman, who gave me permission to reproduce her words:

> My grandfather used to dance as the brother of Sasquatch. We used to go to Happy Camp [in Northern California] and go down the volcanic vents and dance. When he was dancing with his brother he would face me and my grandfather, then turn his back and was 8 feet tall with long hairy arms.

From information provided by members of her tribe, the Yurok Indians of northern California, this shaman had learned that a "family" of sasquatch occupied a territory ranging towards the Northwest all the way to the Yukon.

Sometimes the real and the spiritual worlds coalesce. However, the elders are rather suspicious of those who are strangers to their universe and who often remain quite insensitive to these intercalations. Do they hear the sage who speaks of another time, of profound change, perhaps of hope after chaos, the reversal of the time stream or space itself?

> When the storms blow up the river from the ocean, Grandmother Sasquatch would reverse the flow of the river to return the waters to their calmness...

added the lady shaman.

Chapter 52
Concluding Remarks

The image of the wild man that sticks to my imagination is that of sasquatch. It is inextricably liked to an area, the Pacific Northwest, where lives another being, tall, aromatic, soft to the touch: a magnificent tree, the cedar of the rain forest. The red cedar *(Thuja plicata)* grows up to 230 feet (70 m) tall and may live up to 1,000 years. Another species, the yellow cedar *(Chamaecyparis nootkatensis)* reaches 150 feet (45 m) and may also live for a millennium if spared by the logger's saw.

Both cedar species are used by American Indian sculptors to carve the masks that represent sasquatch. Cedar and its cousins, pine, hemlock and Douglas fir were the mainstay of a wood-based technology: baskets and hats were woven from roots; waterproof clothes made from bark; houses and canoes from wood. Tree products provided the Indians with most of the necessities of life.[1]

Though sasquatch might venture into the sagebrush environment around Omak, the forest is its true kingdom. It roams over the Cascade mountains, the wooded shores of the Columbia, the islands and fjords of Puget Sound and of the coast of British Columbia.

Untold legends remain hidden in this majestic and mysterious scenery, waiting to be deciphered! Before starting on the track of the wild man, I had read a number of sasquatch stories; in one of them one of the protagonists had said: "Only new countries beget strange beasts"[2] A trick to excite the reader's curiosity! The author, Edward Hoagland, introduced the mythical creature:

> But when they [a group of old-time fortune seekers] reached the rim of the little chasm, a mustardy, gingery, joe-pye—weedlike smell did announce the presence of Sasquatch. They heard its peacock or parroty scream, saw its baboon muzzle and bearish set of teeth, its body haired

---
1. See Hillary Stewart's book *Cedar*.
2. Edward Hoagland, *Seven Rivers West*, p. 300.

all over like a shaggy pony's, its hands like a giant man's or an ominous ape's, its forehead and face more spiritual than intelligent but more human than animal. Its body, although dark and strange, looked wholly palpable as it bounded out of the cave, using its fists as forefeet...[3]

There was some information about bigfoot's behavior:

Why are there no Bigfoots outside in the winter? Because Bigfoots hibernate too, but they don't ever die, so they don't get hungry.[4]

The author was rather vague about bigfoot's diet. One of the protagonists related how he had seen bigfoot near a river:

There was a bear there and there was me there and the Bigfoot was there and the salmon running by, but I don't know if he was there to eat the salmon or enjoy the sunshine, because pretty soon he ran away from the bear. That answers you the question of who is more scared of a bear— a Bigfoot or a man. A Bigfoot don't have a gun, you know.[5]

One also learned that Bigfoot had a taste for Indian women: "he's not a bear or an Indian. He's not interested in bears and he won't breed with a sow bear, but he will breed with an Indian. May be he'll steal a girl and afterwards he lets her run home again"[6]

The reader was held in suspense, impatiently awaiting the appearance of the mysterious creature. After a while, it finally showed up, only to disappear and soon reappear again, a rather unnerving experience:

---

3. Hoagland, op.cit. p. 301.
4. Hoagland, op.cit. p. 90.
5. Hoagland, op. cit. p. 123.
6. Hoagland, op. cit. p. 121.

there the beast was. Simio-ursine but also quite like a Brush Man, it sat on the sandy floor of the little chasm, catching and crushing deerflies in its fingers and dipping up water from the edge of the pond and delicately sucking the fins off a shiner, when it happened to catch one.[7]

As to bigfoot's psyche, it could best be understood in comparison to that of the Indian: "But even if its mind was of human calibration...the Bigfoot's was surely an Indian's mind" concluded Hoagland.

Hoagland's novel was best suited for readers fond of epic tales, unfolding in glorious and sometimes foreboding scenery, described in all its splendor. The translator added to the French edition an extensive glossary of terms relative to the flora and fauna of North America, explaining, for example, that kinnikinnick is also the bearberry *(Arctostaphylos uva-ursi)*. Somewhere in the story two buxom and lightly clad Indian beauties are smoking a pipe of kinnikinnick, a plant unknown in Europe. The book is full of such examples. The author lavishly describes the exuberant vegetation, the abundance of wildlife, charismatic large mammals, rodents and birds of all sizes, insects, and so forth. Grass, trees, streams, mountains, clouds and wind surround the protagonists and the local populations. Hoagland's text is a homage to North American nature, North American Indians and the heroes of the story, a quartet of adventurers typical of the pioneers who created modern America.

For the first time, a popular and lively novel spoke of bigfoot. Hoagland was a master of the intricate plot of a picaresque adventure story and brought to life the characters of his fallen heroes, a small group of frontiersmen, including an Indian woman, eager to make their fortune by capturing bigfoot. A clearly impossible task, but the creature haunts its hunters, much like Moby Dick bewitched Captain Ahab.

A brook fell in stair-steps of splashed rock into a pool with bulrushes growing in it and cattails waving hospitably in the breeze on the bank and water striders dashing on the surface

---

7. Hoagland, op. cit. p. 294.

and long-soled Bigfoot tracks ennobling the mud alongside as naturally as the humbler prints of the rock rabbits that had come down to drink.[8]

A clever way to draw the reader's attention to the depth of the footprint; at the time, I wrote in the margin: "The footprint—whether in outline or deeply engraved in the ground—exists within a multimillennial context."[9]

My budding interest in sasquatch and wild men could well have faded away without the stimulus provided by such a captivating novel. I had already read some of Bernard Heuvelmans writings, as well as the works of English and American scientists. It is well worth emphasizing how strongly the interest expressed by a few scientists in the study of wild men stimulates the imagination of the public.

Nevertheless, I might easily have tired of the flood of popular, sensational articles, splashed with fantastic imagery of the Loch Ness monster, or the yeti. The American novels were mostly in the genres of romances or detective stories, with a strong tinge of social Darwinism: the hero, be he a cowboy, detective or journalist, outsmarted the greedy bigfoot hunters, ready to commit any crime for fame and money. The heroine, alluring, bold and practical, was there to help the equally practical champion of law and order. The handsome couple overwhelms the evil hunters, captures bigfoot and unites in happy matrimony. Hence, the most active genes, within athletic bodies and endowed with practical minds, enjoy a definite advantage compared to those less well equipped to handle the struggle for existence.

Fortunately, there are also some excellent "origins" novels, such

---

8. Hoagland, op. cit. p. 288.

9. Marc Yvard, then a student and the son of one of my fellow English teachers expressed some pertinent comments in an essay entitled: "The Cryptozoological Footprint" (*L'Empreinte cryptozoologique*, 1992). For example, " The footprint of this unknown creature resembles the spilling of dream into reality described by Nerval," and, "Anyone feeling trapped by the invasive reality, be it tangible or virtual, can only wish when discovering unexplained footprints to become a fisher of dreams or a hunter of chimeras."

as Jean Auel's multivolume prehistoric saga *Earth's Children,* remarkable for the author's extensive paleontological erudition.

Today's interest in human origins is well served by carefully documented but highly readable works of fiction such as John Darton's, *Neanderthal* (1996), or Greg Bear's science-fiction novels, *Darwins' Radio* (1999; Nebula Prize, 2000) and *Darwin's Children* (2003).[10]

The quality of the documentation found in works of fiction has over the past years reached a very high level. For example, while maintaining his readers' interest, Swiss writer Jean-Marc Pasquet demonstrated a remarkable knowledge of Pacific Northwest mythology. After leaving Switzerland, his protagonist finds himself caught in a world of ruthless adventurers. He is sheltered by a couple who lead him through a holistic initiation; his masters are a shaman and a female sasquatch named Qâ; hence the title of the novel *Le Don de Qâ,* (2001; *Qâ's Gift*). As to Qâ, who is she? A sadly and bestially ugly woman, or an unbearably human primate?

During a book-signing session, I met Jean-Marc Pasquet and asked him how long he had lived on the west coast of the USA. He admitted that he had never been there before writing *Le Don de Qâ,* having worked from books, and information gathered through letters and e-mails with Americans and Indians. Pasquet was fascinated by what he had learned; he has since visited the locale of his novel. "I felt like I had to catch up," he said.

Sometimes, tired of the mediocrity of some novels, I preferred to reread Eric Buffetaut's small volume on the history of paleontology, a series of scientific adventures starring Cuvier, Lamark, and Darwin. I also delved with great pleasure into the works of neo-Darwinist Stephen Jay Gould. His work, like that of Carl Sagan, showcases the talent and fertile imagination of a brilliant essayist.

Gradually, however, I distanced myself from those Darwinian explanations, which seek an answer to every evolutionary problem. There soon arises a tendency for post-hoc justification, within an

---

10. By the way, I wish to point out the pertinence of the view expressed by one of Bear's characters in *Darwin's Radio,* to the effect that while everyone thinks the Indians are ignorant, in fact it is simply that the Indians and the whites are concerned with different things.

evolutionary context, of every phenomenon of biological change. It all ends up being for the best in the best of all possible worlds.

Sometimes then some doubt lingers in my mind, and in spite of the admirable work of Darwin and his successors, my curiosity remains unsatisfied. The basic thirst for understanding that motivates my investigation of the wild men brings about surprising discoveries. For example, on the basis of the Bible, the Italian Attilio Mordini contends that the yeti, a descendant of the race of Cain, is a "progression" of man towards the beast. "Cain withdrew from the face of Yawveh and dwelled in the land of Nod, east of Eden" (Genesis, IV, 16). East of Eden is at the edge of India and the Himalayas. For Mordini, "between the Yeti and the rest of humanity, there lies the great flood."[11] The depth of the chasm that separates us from the yeti is, of course, on the scale of the Biblical cataclysm.

On the theatrical side, I keep a most pleasant memory of a Jon Klein's play, *Betty the Yeti*, which some friends invited me to see in Seattle on a Sunday night in 1994. It had an ecological theme, and featured loggers, wearing plaid shirts and wide suspenders, working for a ruthless corporation. Also starring were an owl, whose presence was a sign of the desirable ecological status of the forest, as well as bigfoot, another indicator of the health of the wilderness.

On a stage decorated with cardboard trees, representing the Willamette National Forest of Oregon, the actors delivered rather lofty-sounding lines in a performance somewhat reminiscent of a high school production. It was all in support of a worthy cause, but I can't forget the occasionally amusing (sometimes accidentally so) moments of *Betty the Yeti*.

Fortunately, good luck sometimes leads to the discovery of most interesting gems where they are not expected. For example, a most stimulating book, *My Quest for the Yeti*, was published in 1998 by the famous alpinist Reinhold Messner. Messner climbed all 14 Himalayan peaks over 26,250 feet (8,000 m); in 1986, he reached the summit of Everest without oxygen tanks.

I cannot resist quoting a few lines of his work:

Without Wilderness There is No Yeti. Thus, the survival of

---

11. Attilio Mordini, *Le Mystère du Yeti*, p. 61.

the Yeti myth is dependent on the survival of the last wilderness—areas so undeveloped that even the local population cannot conclusively confirm or refute the equation of the brown bear with the Yeti. There is much more behind our thirst for monsters than curiosity or escapism. There is the fear that the earth is losing the last regions where myths can flourish.[12]

Among the dozen of works that inspired my search for the wild man, as well as among the plethora of reports and opinions I gathered over the years, I somewhat regretfully had to make a choice. I must now think about some concluding words.

Coming to the end, I wish to speak of a man I did not have the occasion of meeting, Jordi Magraner. Friends met him and told me about him; I saw him on the French–German TV network Arte, the only channel that refrained from ridiculing his work on the sasquatch and its relatives.[13] The documentary featured the Chitral region, in northern Pakistan, an area where the peaks commonly reach over 20,000 feet (6,000 m). This region, formerly known as the Lesser Kachgar, abuts the Chinese Kachgar, the permanent home of the wild men, according to Porchnev. One sees very few tourists there.

The sedentary Chitralis live in the valleys; small groups of nomadic Gujar herders dwell on the mountain slopes. The higher levels are practically uninhabited. In Chitrali, the wild men are called *jangali mosh,* "men of the forest," or "wild men," the term *almasty*, "he who eats a lot," is rarely used; the most common term is *barmanou,* "the strong, or muscular man," probably a Gujari word.

Over 19 months, PhD zoologist Jordi Magraner, crisscrossed the area, accompanied by brothers Erik and Yannick L'Homme and a pair of satchel-carrying malamute dogs. The Arte documentary

---

12. Reinhold Messner, *My Quest for the Yeti,* Pan Books, London, 2001, p. 13.
13. In December 2001, The Arte channel broadcast "Cryptopuzzle: What Drives Cryptozoologists?" directed by Jacques Mitsch, technical advisor Benoit Grison. Participants included Eric Buffetaut, Marie-Jeanne Koffmann, Yves Coppens (who defined the almasty as a large primate), Dmitri Bayanov and Pascal Tassy.

lavishly documents the mountains and forest of the Hindu Kush which they explored. Success in such an expedition depends critically on mountaineering skills and familiarity with local languages. Magraner's enquiry protocol followed five steps and was based on three essential elements:

—Spontaneous reports by witnesses;
—A 63-point questionnaire;
—Sketches: those created by the witnesses as well as those selected among a choice of suggestions.

The questionnaire was based on an analysis of the frozen Minnesota wild man, named *Homo pongoides* by Heuvelmans, which presented the most detailed set of characteristics of such creatures.

The 27 testimonies collected on the wild men of the Chitral actually do correspond to Heuvelmans' pongoid man. While Magraner's report requires familiarization with some technical vocabulary, his conclusions, although nuanced, are of great interest:

The synthetic result of these reports does not enable us to support the hypothesis that it is of purely mythical origin without any natural grounds. Both anatomical data of fossil men and the prehistory of central Asia do not contradict the existence of prehistoric populations in high altitudes of Pamir and Hindu Kush. This was true at least until the first millenary BC. The question is whether oral statements could not be relative to survivors of these very late prehistoric populations, already described by the Iron Age population of Sakas, the first shepherds of the Bactriane."[14]

It thus seems to be a historical truth that people with a very ancient cultural tradition settled in the Pamir at the end of the Neolithic Age and lived there at least as late as 1,000 BC. Who were they? Jordi Magraner was going to find out in his third expedition. Unfortunately, he was killed in Pakistan on August 2, 2002, as were

---

14. Jordi Magraner. In: Oral Statements concerning living unknown hominids: analysis, criticism and implications for language origins. (Abstract)
<http://www.bigfootencounters.com/biology/jordi.htm>

his guide and a 12-year-old assistant. They were found with their throats slit.

In his own words, Magraner was hoping to be able to carry on a "free and unconstrained investigation" in North Pakistan. Funded mainly by benefactors and contributions from individual supporters, he also worked as a guide for aid agencies and collected plants for pharmaceutical laboratories. He rented a small hut, with a fenced yard for the dogs. The owners were small-scale farmers. The barmanou was said to have come and helped itself to some ears of the corn that they grew.

Looking for witnesses, Magraner went from valley to valley, climbing over steep mountains. Uncontrolled clear cuts had shrunk the forest; the barmanou habitat was reduced to a zone between 6,500 and 13,000 feet (2,000 and 4,000 m). Magraner was happy with little, and adapted to sometimes harsh living conditions.

I have often looked at the map of central Asia and the surrounding countries—Bhutan, Nepal, Tibet, Pakistan, China—I was mesmerized by the Hindu Kush. I would start daydreaming simply upon hearing the name. I was immediately a silent admirer of Jordi Magraner. Some of my friends shared these feelings and considered Magraner as a proxy for their own curiosity, an extension of their interest in the wild man. The field investigator was in a way the ambassador of our enthusiasms. Unfortunately, our sympathy could do nothing to prevent his murder.

Jordi Magraner embodied in the East, the qualities of the researchers that I had met in the West. Like them, he lived simply, connected with the land, the flora and the fauna with scientific rigor. He learned the local languages and their dialects and was familiar with the traditions and the imagination of his informers. He believed that within a few months he was likely to be able to capture the wild man on film.

May he rest in peace. I am sure that other seekers/investigators will follow in his footsteps. Like him they will try to prove that a different branch of humanity has survived. Other researchers will pursue a similar quest in the Pacific Northwest. Some day, an American Indian will tell them: "Sasquatch is the spirit of our people."

These words will, I hope, leave a lasting imprint.

Additional Photographs from Author's File

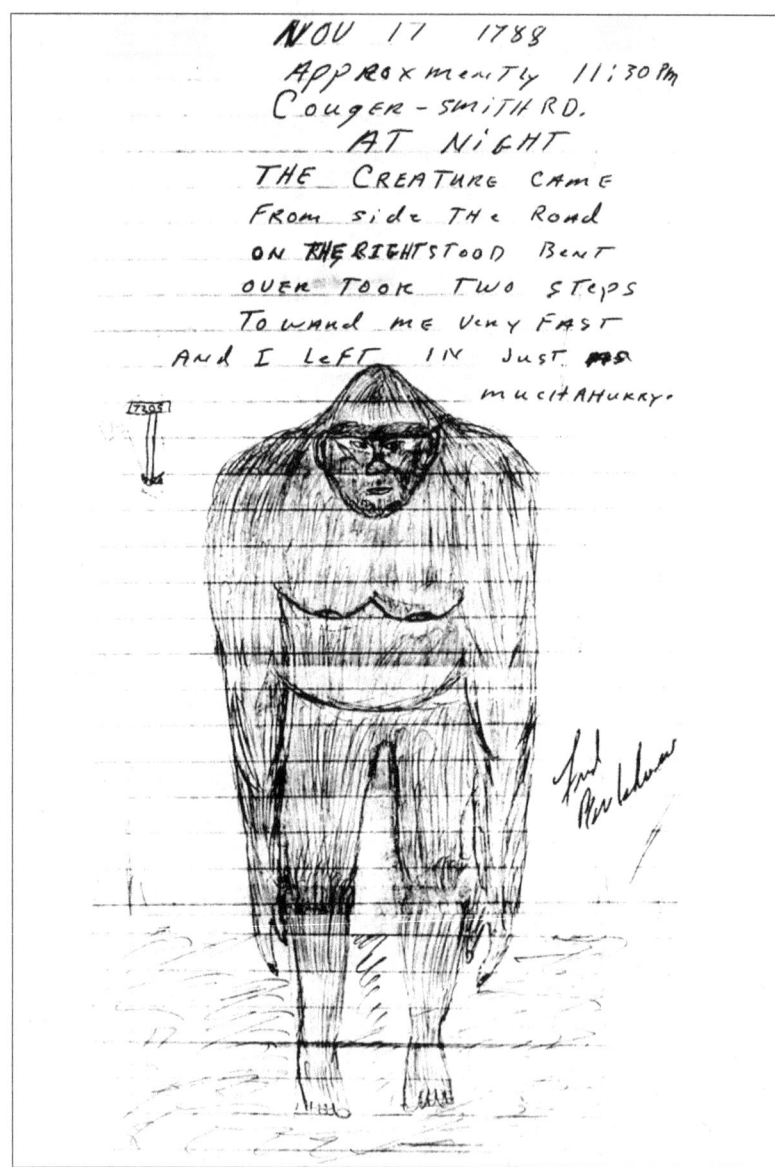

Bigfoot encounter: Text and drawing by Fred Bradshaw.

(Top) Almasty by Marie-Jeanne Koffmann.
(Lower) Cave painting, Neanderthal portrait, Isturitz Pyrénées, Atlantiques, France. Note the hairiness, the low forehead, upturned nose and weak chin.

(Top) A Barmanu after Jordi Magraner, based on the report of a Pakistani shepherd (September 1977).
(Lower) Wild men of Shennongjia, China, (etching, 1455 AD).

Wild man: Church of Beaulieu sur Dordogne, Corrèze France (ninth century). The tree that he is holding, like Hercules' club, is a symbol of his mastery of the wild forest.

A "showman" and his "beast." Church sculpture: Notre-Dame, Semur-en-Auxois (Côte d'Or, France), twelfth century.

Wild man holding a staff and shield. Church sculpture: Ambierle (Loire, France), sixteenth Century.

Wild woman and offspring. Church sculpture: Ambierle (Loire, France), sixteenth century. Note the bare knees.

Sculpture of a bear being slain by a knight (eleventh to twelfth century; Andlau Abbey, Bas-Rhin, France). This scene shows how the church was putting down pagan rituals, such as the cult of the bear.

Bal des Ardents (Ball of Burning Men, fourteenth century). The costumes of King Charles VI and courtiers, disguised as wild men, caught fire during a ball. (See chapter 39, page 293.)

Game of the Bear in Prats-de-Mollo (Pyrénées, France). The Feast of the Bear takes place every year. The bear—a villager in disguise—comes down from the hills to terrorize the people.

A man comparing his size 14 boot to a cast of a footprint found in Washington State in the 1970s. The cast measured 17 inches (43.2 cm) long.

Bigfoot model created by the Belgian artist Emmanuel Janssens (March 2000).

(Above) The Colville Confederated Tribes erected this metal sculpture created by Virgil "Smoker" Marchand in honor of bigfoot. The sculpture is located at Disautel Pass on the Colville reservation, Washington State (see references: chapter 12, page 95; and chapter 15, page 106).

(Left) Ed Fusch (left) and author in front of sculpture by Smoker Marchand, July 2007.

Left to right: Michael Noonan, Todd Deery, Peter Byrne, Deborah Wolman, Dr. Jean-Paul Debenat, and Loren Coleman. The photograph was taken at the BFRP (Bigfoot Research Project) in Mount Hood, OR, on April 14, 1994.

Bibliography

Books

Alley, Robert J. 2003. *Raincoast Sasquatch.* Surrey, B.C., Canada: Hancock House Publishers.
Arrou-Vignod, Jean-Philippe. 1997. *L'Homme du cinquième jour.* Paris, France: Gallimard.
Ashwell, Reg. 1992. *Indian Tribes of the Northwest.* Surrey, B.C., Canada: Hancock House Publishers.
Barloy, Jean-Jacques. 1975. *Serpent de mer et monstres aquatiques.* Geneva, Switzerland: Editions Famot.
Bayanov, Dmitri. 1996. *In the Footsteps of the Russian Snowman.* Moscow, Russia: Crypto-Logos.
———. 1997. *America's Bigfoot: Fact, Not Fiction.* Moscow, Russia: Crypto-Logos.
Bear, Greg. 1999. *Darwin's Radio.* New York: Del Rey Books.
———. 2003. *Darwin's Children.* New York: Del Rey Books.
Beck, Mary G. 1993. *Potlatch.* Seattle: Alaska Northwest Books.
Bernard, Alain. 1998. *La Cuisine Préhistorique.* Périgueux, France: Editions Fanlac.
Bindernagel, John A. 1998. *North America's Great Ape: the Sasquatch.* Courtenay, B.C., Canada: Beachcomber Books.
Bly, Robert. 1990. *Iron John.* Reading, MA: Addison-Wesley Publishing Co.
Boas, Franz. 1955. *Primitive Art.* New York; Dover Publications.
———. 1995. *A Wealth of Thought.* Seattle: University of Washington Press.
——— and James Teit. 1985. *Coeur d'Alene, Flathead and Okanogan Indians.* Fairfield, WA: Ye Galleon Press.
de Bonis, Louis. 1999. *La famille de l'homme.* Paris, France: Editions Belin.
Boston, John. 1993. *Naked Came the Sasquatch.* Lake Geneva, WI: TSR Inc.
Bournaud, Michel. 1997. *Contes et Légendes de l'Ours.* St Claude de Diray, France: Editions Hesse.

Bradbury, Ray. 1950. *The Martian Chronicles.* New York: Doubleday.
Bramly, Serge. 1992. *Terre Sacrée.* Paris, France: Editions Albin Michel.
Briley, Ann. 1986. *Lonely Pedestrian: Francis Marion Streamer.* Fairfield, WA: Ye Galleon Press.
Brown, Vinson. 1993. *Native Americans of the Pacific Coast.* Happy Camp, CA: Naturegraph Publishers.
Bruchac, Joseph. 1993. *The Native American Sweat Lodge.* Freedom, CA: The Crossing Press.
de Bruycker, Daniel. 2001. *Silex.* Arles, France: Editions Babel/Actes Sud.
Byrne, Peter. 1976. *The Search for Bigfoot: Monster, Myth or Man?* New-York: Pocket Books.
Campbell, Joseph. 1990. *The Flight of the Gander.* New-York: Harper Perennial.
Chevalier, Jean and Alain Gheerbrandt. 1973. *Dictionnaire des symboles.* Paris, France: Editions Seghers.
Clark, Ella E. 1953. *Indian Legends of the Pacific Northwest.* Berkeley, CA: University of California Press.
Clébert, Jean-Paul. 1971. *Dictionnaire dy Symbolisme animal.* Paris, France: Albin Michel.
Clostermann, Pierre. 2005. *Une Vie pas comme les autres.* Paris, France: Flammarion.
Colhene, Terri. 1990. *Clamshell Boy: A Makah Legend.* Vero Beach, FL: Watermill Press.
Combes, C., Y. Coppens, O.Coudart, J.-J. Dutour, A., Hublin, B. Langaney, and B. Vandermeersch. 1994. *L'Homme: Origine et Destinée.* Paris, France: Editions Errance.
Coppens, Yves. 1999. *Le Genou de Lucy.* Paris, France: Editions Odile Jacob.
Crowe, Ray. 2000. *The Shaman from the East.* Hillsboro, OR: Western Bigfoot Publishing.
Darnton, John. 1998. *Neandertal.* Paris, France: Editions J'ai Lu.
Delfe, Gérard. 2004. *Le Dieu Coyote.* Paris, France: Editions Phébus.
Deloison, Yvette. 2004. *Préhistoire du piéton.* Paris, France: Editions Plon.

Diamond, Jared. 1992. *The Rise and Fall of the Third Chimpanzee.* London, UK: Vintage Publisher.
Dibble, Barney. 1980. *Pan.* New York: Leisure Books.
Dietrich, William, 1995. *Northwest Passage.* New-York: Simon and Shuster.
Doyle, Paul. 1992. *Nioka Bride of Bigfoot.* Seattle: Daily Planet Press.
Downing, Alfred. 1987. *The Region of the Upper Columbia and How I Saw It.* Fairfield, WA: Ye Galleon Press (reprint of 1881 edition).
Drew, Leslie. 1992. *Haida.* Surrey, B.C., Canada: Hancock House Publishers.
Dupont, Jean-Claude. 1992. *Légendes Amérindiennes.* Québec, Canada: Editions Jean-Claude Dupont.
———. 1993. *Les Amérindiens au Québec.* Québec, Canada: Editions Jean-Claude Dupont.
Durrell, Gerald and Lee. 1993. *The Amateur Naturalist.* London, UK: Dorling Kindersley.
Eliade, Mircea. 1983. *Le chamanisme.* Paris, France: Editions Payot.
Ellul, Jacques. 1990. *La Technologie ou l'Enjeu du Siècle.* Paris, France: Economica.
d'Esme, Jean. 1928. *Les Dieux Rouges.* Paris, France: Editions Plon.
Fusch, Ed. 1992. *They Walked Among Us: Scweneyti and the Sticks Indians of the Colvilles.* Riverside, WA: Ed Fuchs (P.O. Box 47, Riverside, WA 98849).
Fredin, Adeline and Tom Crawford. 1978. *Traditional Teachings of the Colville Confederated Tribes.* (No further details).
Galdikas, Biruté M.F. 1995. *Reflections of Eden.* London, UK: Victor Gollancz.
Gérard-Landry, Chantal. 1995. *Hopi, Peuple de Paix et d'Harmonie.* Paris, France: Albin Michel.
Geronimo (S.M. Barrett, Ed.). 1971. *Geronimo, His Own Story.* New York: Ballantine Books.
Gill, Sam D. and Irene F. Sullivan. 1992. *Dictionary of Native American Mythology.* Oxford, UK: Oxford University Press.
Gloss, Molly. 2000. *Wild Life.* Boston: Mariner Books.

Glickman, J. 1998. *Toward a Resolution of the Bigfoot Phenomenon.* Hood River, OR: North American Science Institute.
Goodall, Jane. 1997. *The Chimpanzee Family Book.* New York: North-South Books.
Gordon, David G. 1992. *Field Guide to the Sasquatch.* Seattle: Sasquatch Books.
Gould, Stephen J. 1987. *An Urchin in the Storm.* New York: W.W.Norton Publishers.
———. 1994. *Un hérisson dans la tempête.* Paris, France: Grasset, Editor. (Translation of *An Urchin in the Storm.*)
———. 1999. *Rocks of Ages: Science and Religion in the Fullness of Life.* New York: Ballantine Books.
———. 2000. *Et Dieu dit: "que Darwin soit": science et religion, enfin la paix?* Paris, France: Editions du Seuil. (Translation of *Rocks of Ages.*)
———. 2000. *The Lying Stones of Marrakech.* London, UK: Vintage Publisher.
Green, John. 1970. *Year of the Sasquatch.* Agassiz, B.C., Canada: Cheam Publishing.
———. 1980. *On the Tracks of the Sasquatch* (Book 1 and Book 2). Harrison Hot Springs, B.C. Canada: Cheam Publishing.
Griaule, Marcel and Germaine Dieterlen. 1991. *Le renard pâle (Tome 1).* Paris, France: Institut d'Ethnologie.
Gribbin, John and Michael White. 2000. *Darwin: a Life in Science.* London, UK: Simon and Schuster.
Guillaud, Lauric. 2003. *La Terreur et le Sacré.* Paris, France: M. Houdiard, Editeur.
Guthrie, Woody. 1943. *Bound for Glory.* New York: E.P. Dutton.
Halpin, Marjorie and Michael M. Ames, eds. 1980. *Manlike Monsters on Trial: Early Records and Modern Evidence.* Vancouver, B.C. Canada: University of B.C. Press.
Harris, Christie. 1975. *Sky Man on the Totem Pole?* New York: Atheneum Press.
Hausman, Gérald. 1996. *Bestiaire des Indiens d'Amérique.* Paris, France: La Table Ronde.
Heath-Stubbs, John. 1988. *Collected Poems 1943–1987.* Manchester, UK: Carcanet.
Hélias, Pierre-Jakez. 1975. *Le Cheval d'orgueil.* Paris, France: Plon,

Editeur. (Translated as: *The Horse of Pride*. 1980. New Haven, CT: Yale University Press.)

Herrick, Robert. 1856. *Hesperides: or the works both humane and divine of Robert Herrick, Esq. Vol. II*. Boston: Little Brown & Co.

Heuvelmans, Bernard. 1955. *Sur la piste des bêtes ignorées*. Paris, France: Plon, Editeur. (Translated as: *On the Track of Unknown Animals*. 1995. London, UK: Kegan Paul International.)

———. 1958. *Dans le sillage des monstres marins*. Paris, France: Plon, Editeur.

———. 1965. *Le Grand Serpent-de-Mer*. Paris, France: Plon, Editeur. (These last two books by Heuvelmans have appeared in English combined as: *In the Wake of the Sea-Serpents*. 1968. NewYork: Hill and Wang.)

———. 1978. *Les derniers dragons d'Afrique*. Paris, France: Plon, Editeur.

———. 1980. *Les Bêtes Humaines d'Afrique*. Paris, France: Plon, Editeur.

———. 2006. *The Kraken and the Colossal Octopus*. London, UK: Kegan Paul.

———, with Boris Porchnev, 1974. *L'Homme de Néanderthal est toujours vivant*. Paris, France: Plon, Editeur.

Hilderbrand, Sandra (compiled by). 1991. *Treasure in the Okinagan–Oroville Area History (Vol.1)*. Colville, WA: Colophon Press.

Hilbert, Vi. 1985. *Haboo*. Seattle: University of Washington Press.

Hoagland, Edward. 1986. *Seven Rivers West*. New York: Summit Books.

Holm, Bill. 1993. *Northwest Coast Indian Art*. Seattle: University of Washington Press.

Hoyt, Richard. 1993. *Bigfoot*. New York: Tor Publishing.

Hulkrantz, Ake. 1993. *Religions des Indiens d'Amérique*. Aix-en-Provence, France: Editions Le Mail.

Humphrey, Nicholas. 1993. *The Inner Eye*. London, UK: Vintage Publishing.

Hunn, Eugene S. 1990. *Nch'i-Wána "The Big River."* Seattle: University of Washington Press.

Hunter, Don and René Dahinden. 1973. *Sasquatch: The Search for*

North America's Incredible Creature. Toronto, Canada: McLellan and Stewart.

Joisten, Alice and Christian Abry. 1995. *Etres fantastiques des Alpes.* Paris, France: Editions Entente.

Joly, Eric and Pierre Affre. 1995. *Les Monstres sont vivants.* Paris, France: Grasset, Editeur.

Jonaitis, Aldona, Ed. 1994. *A Wealth of Thought—Franz Boas on Native American Art.* Seattle: University of Washington Press.

Jung, Carl Gustav. 1968. *Man and His Symbols.* Topeka, KS: Topeka Bindery.

Kipp, Neil. 1980. *Pre-Columbian History of the Red Man.* Browning, ME: Blackfeet Heritage Program.

Knight, Rainy. 1991. *Critter.* Rochester, WA: Gorham Printing.

Krantz, Grover. 1992. *Big Footprints: A Scientific Enquiry into the Reality of Sasquatch.* Boulder, CO: Johnston Books. Revised Edition, 1999. Surrey, B.C., Canada: Hancock Publishers.

Lall, Kesar. 1988. *Lore and Legend of the Yeti.* Thamel, Khatmandu, Nepal: Pilgrims Book House.

LeBlond, Paul and E.L. Bousfield. 1995. *Cadborosaurus.* Victoria, B.C., Canada: Horsdal and Schubart Publishers.

Leenhardt, Maurice. 1998. *Le Pilou, moment culminant de la société.* Nouméa, Nouvelle Calédonie: Editions Grain de Sable.

LeNoel, Christian. 2000. *La Race Oubliée, Tome 1.* Toulon, France: Editions Cheminements.

———. 2005. *La Race Oubliée, Tome 2.* Toulon, France: Editions Cheminements.

Leroy, Patrick. 2004. *Le Yéti et autres bipèdes poilus....* Cazoul-les Béziers, France: Editions du Mont.

Lévi-Strauss, Claude. 1979. *La Voie des Masques.* Paris, France: Editions Pocket. (Translated in 1982 by Sylvia Modelski as: T*he Way of the Masks.* Seattle: University of Washington Press.)

Lewis, Meriwether and William Clark. 2002. *Journals.* Washington, DC: National Geographic Adventure Classics.

Ley, Willy. 1959. *Exotic Zoology.* New York: Random House.

Lindbergh, Alika. 1958 (under the pseudonym Monique Watteau). *L'Ange à la fourrure.* Paris, France: Plon, Editeur.

———. 1976. *Quand les singes hurleurs se tairont*. Paris, France: Presses de la Cité.

———. 2002. *Testament d'une Fée*. Paris, France: e/dite.

Logan, Nancy R. 1987. *Children of a Lost Spirit*. Mercer Island, WA: Kideko House Books.

Lumley, Henri de. 1998. *L'Homme Premier*. Paris, France: Odile Jacob, Editeur.

Mackin, J.H. and A.S. Cary. 1965. *Origin of Cascade Landscapes*. Olympia, WA: State Printing Plant.

Malaurie, Jean. 2003. *L'Allée des Baleines*. Paris, France: Mille et Une Nuits.

Markotic, Vladimir. 1984. *The Sasquatch and Other Unknown Hominoids*. Calgary, AB, Canada: Western Publishers.

Matthiessen, P. 1978. *The Snow Leopard*. London, UK: Penguin Books.

Mayr, Ernst. 2002. *What Evolution Is*. London, UK: Phoenix.

Mercier, Mario. 1987. *Chamanisme et chamans*. St-Jean-de-Braye, France: Editions Dangles.

Merle, Robert. 1991. *Le Propre de l'Homme*. Paris, France: Livre de Poche.

Messner, Reinhold. 2000. *Le Yéti: Du mythe à la réalité*. Paris, France: Glénat, Editeur. (Translated by Peter Constantine as: *My Quest for the Yeti*. London, UK: Pan Books, 2001.)

Millar, Ronald. 1997. *The Green Man*. Sanford, UK: S.B. Publications.

Mooney, James. 1973. *The Ghost-Dance Religion and Wounded Knee*. New York: Dover Publications.

Mordini, Attilio. 1987. *Le Mystère du Yéti*. Puiseaux, France: Pardès, Editeur.

Mourning Dove. 1990. *Coyote Stories*. Lincoln, NE:University of Nebraska Press.

Murphy, Christopher L., with John Green and Thomas Steenberg. 2004. *Meet the Sasquatch*. Surrey, B.C., Canada: Hancock House Publishers.

Napier, John R. 1972. *The Yeti and Sasquatch in Myth and Reality*. London, UK: Jonathan Cape.

Newton, Michael. 2007. *Florida's Unexpected Wildlife*. Gainesville, FL: University Press of Florida.

Nolan, Richard D. 1996. *Sur les traces du Yéti et autres créautre clandestines.* Toulon, France: Plein Sud.
Ossendowski, Ferdinand. 1969. *Bêtes, hommes et dieux.* Paris, France: Editions J'ai Lu.
Page, George. 1999. *The Singing Gorilla.* London, UK: Headline Publishers.
Pasquet, Jean-Marc. 2001. *Le Don de Qâ.* Paris, France: Editions Jean-Claude Lattès.
Patterson, Roger. 1966. *Do Abominable Snowmen of America Really Exist?* Yakima, WA: Franklin Press. (Republished in 1967, Yakima, WA: Northwest Research Association; in 1996, New Westminster, B.C., Canada: Pyramid Publications; in 2005 under the new title *The Bigfoot Film Controversy* , Surrey, B.C., Canada: Hancock House Publishers.)
Pernaud-Orliac, Jacques. 1997. *Petit Guide de la Préhistoire.* Paris, France: Editions du Seuil.
Pichette, Pierre, 1974. *Coyote Tales of the Montana Salish.* Rapid City, SD: Tipi Shop Inc.
Pichon, Jean-Charles. 1971. *Histoires des Mythes.* Paris, France: Petite Bibliothèque Payot.
———. 1972. *Les Dieux Etrangres.* Paris, France: Editions Payot. Republished in 2000 as *Le Cri Articulé:* e/dite.
———. 1986. *L'Homme et les Dieux.* Sainte-Ruffine, France: Editions Maisonneuve.
——— 2000. *Litanies des Dieux Morts.* Paris, France: e/dite.
Picq, Pascal. 1999. *Les Origines de l'Homme.* Paris, France: Tallandier.
Pyle, Robert M. 1995. *Where Bigfoot Walks.* Boston: Houghton Mifflin Co.
Radin, Paul. 1988. *The Trickster: A Study in American Indian Mythology.* New York: Shocken Books.
Ray, Verne F. 1933. *The Sanpoil and Nespelem.* Seattle: University of Washington Press.
Reeves, H., J. de Rosnay, Y. Coppens and D. Simonnet. 1996. *La plus belle histoire du monde.* Paris, France: Editions du Seuil.
Reid, Bill. 1989. *Le Dit du Corbeau.* Paris, France: Atelier Alpha Bleu.
——— and Robert Bringhurst. 1996. *The Raven Steals the Light.* Boston: Shambhala Publications.

Rimbaud, Arthur. 1962. *Poèmes*. Paris, France: Livre de Poche.
Roche, Jean. 2000. *Sauvages et Velus*. Chambéry, France: Editions Exergue.
Rockwell, David. 1991. *Giving Voice to Bear*. Niwot, CO: Roberts Rinehart.
Roosevelt, Theodore. 1893. *The Wilderness Hunter*. New York: Putnam and Sons.
Rose, Steven. 1997. *Lifelines*. London, UK: Penguin Books.
Rosny, J.H. 1985. *La Guerre du Feu*. Paris, France: Livre de Poche. (Translated as: *The Quest for Fire*. 1982. New York: Ballantine.)
Ruby, Robert H. and John A. Brown. 1993. *Indian Slavery in the Pacific Northwest*. Spokane, WA: The Arthur H. Clarke and Co.
Ruby, Robert H. and John A. Brown. 1982. *The Spokane Indians*. Norman, OK: University of Oklahoma Press.
Sagan, Carl. 1977. *The Dragons of Eden*. New York: Random House.
de Saint-Exupéry, Antoine. 1943. *Le Petit Prince*. Paris, France: Gallimard.
Sakurai, Atsushi. 1984. *Salmon*. New York: Alfred Knopf.
Sanderson, Ivan T, 1977. *Abominable Snowman, Legend Come to Life*. New York: Jove/HBJ.
Shackley, Mira. 1986. *Still Living? Yeti, Sasquatch and the Neanderthal Enigma*. New York: Thames and Hudson.
Schaller, George B. 1988. *The Stones of Silence: Journeys in the Himalayas*. Chicago: University of Chicago Press.
———. 1994. *The Last Panda*. Chicago: University of Chicago Press.
Schlick, Mary D. 1994. *Columbia River Basketry*. Seattle: University of Washington Press.
Sharp, Henry S. 1988. *The Transformation of Bigfoot*. Washington, DC: Smithsonian Institution Press.
Smith, J.L.B.1956. *Old Four Legs: The Story of the Coelacanth*. London, UK: Longmans, Green.
———. 1960. *The Search Beneath the Sea: The Story of the Coelacanth*. New York: Henry Holt Publishing.
Smith, Malcolm. 1996. *Bunyips and Bigfoots*. Alexandria, Australia: Millennium Books.

Snyder, Gary. 1991. *He Who Hunted Birds in His Father's Village*. San Francisco: Grey Fox Press.

Sprague, Roderick and Grover Krantz. 1979. *The Scientist Looks at the Sasquatch (Book 2)*. Moscow, ID: University Press of Idaho.

Stewart, Hilary. 1995. *Cedar*. Seattle: University of Washington Press.

Swanton, John R. *Indian Tribes of Washington, Oregon and Idaho*. Fairfield, WA: Ye Galleon Press.

Tassi, Franco. 1990. *Animali a Rischio*. Milano, Italy: Giorgio Mondadori e Associati.

Tattersall, Ian. 1998. *Becoming Human: Evolution and Human Uniqueness*. New York: Harcourt Brace.

Teit, James A., Marian K. Gould, Livingstone Farrand and Herbert J. Spinden. 1969. *Folk-tales of Salishan and Sahaptin Tribes* (edited by Franz Boas). New York: Klaus Reprint Co. (Reprinted from the American Folklore Society, 1917.)

Trafzer, Clifford E. 1992. *The Nez Perce*. New York: Chelsea House Publishers.

Turcan, Robert, 1997. *The Cults of the Roman Empire* (translated by Antonia Nevill). Oxford, UK: Blackwell Publishing.

Vercors. 1999. *Les Animaux Dénaturés*. Paris, France: Livre de Poche. (Translated as: *You Shall Know Them*. 1953. Boston: Little, Brown and Company.)

Versluis, Arthur. 1994. *Amérindiens qui êtes-vous?* St-Martin-le-Vinoux, France: Editions L'Or du Temps.

de Waal, Frans and Frans Lanting. 1999. *Bonobos*. Paris, France: Editions Fayard.

Ward, Peter Douglas. 1992. *On Methuselah's Trail—Living Fossils and the Great Extinctions*. New York: Freeman and Co.

Wasson, Barbara. 1979. *Sasquatch Apparitions*. Bend, OR: B.Wasson, P.O. Box 5551, Bend, OR.

West, Bonnie. 1993. *Bigfoot! He's Still Out There!* Everett, WA: Lowel Printing and Publishing. (Vol. 3 of *The Sasquatch Have Their Ways* series).

Wolff-Quenot, Marie-Josèphe. 1996. *Des Monstres aux Mythes*. Paris, France: Trédaniel, Editeur.

Wyatt, Gary. 1994. *Spirit Faces*. London, UK: Thames and Hudson Ltd.

Yanan, Eileen. 1971. *Coyote and the Colville*. Omak, WA: St.Mary's Mission.
Zhang, Qing. 2005. *Yeren*. Hohhot, Inner Mongolia, China: Yuan Fang Publishing House.

Magazines and Newspaper Articles:

Heuvelmans, B. 1986. "Annotated checklist of Apparently Unknown Animals with which Cryptozoology is concerned." Tucson, AZ: *Cryptozoology*, 5, p. 12.
Batiste, Martine. 2000. "Notre ancêtre prend un coup de vieux." Paris, France: *L'Express*, December 14.
Sanderson, Ivan T. 1959. "The Strange Story of America's Abominable Snowmen." New York: *True*, December.
Scott, P. and R. Rines. 1975. "Naming the Loch Ness Monster." London, UK: *Nature*, Vol. 258, pp 466–468, December 11.
Thireau, Michel. 1987. "Une guerre, une bombe et le plus grand gecko du monde." Paris, France: *L'Univers du Vivant*, March.
Lagrange, Pierre. 2004. "Florès: Yéti y es-tu?" Paris, France: *Sciences et Avenir*, December.
Mamère, Noël. 2001. "La Nouvelle Généalogie ADN." Paris, France: *Sciences et Avenir*, April.

Publication Organizations Mentioned:

BIPEDIA
Editor: François de Sarre
Centre d'Etudes et de Recherches sur la Bipédie Initiale.
32, avenue de Buenos Aires,
06000 Nice, France
Website:
www.pagesperso-orange.fr/initial.bipedalism/bipedia.htm

CRYPTOZOOLOGIA
Editor: Eric Joye
Association Belge d'Etude et de Protection des Animaux Rares.
Square des Latins, 49/4
B-1050 Bruxelles, Belgium
Website: www.cryptozoology.be/

HOMINOLOGIE ET CRYPTOZOOLOGIE
Editor: Christian Le Noël
Association Française de Recherches cryptozoologiques.
20, avenue Maurice Garçon
86240 Ligugé, France
Website: www.pagesperso-orange.fr/darue/cryhom.htm

INSTITUT VIRTUEL DE CRYPTOZOOLOGIE
Site created by Michel Raynal
Website: www.pagesperso-orange.fr/crytozoo/index.htm

THE TRACK RECORD (No longer published)
Ray Crowe, Editor
225 NE 30th Avenue
Hillsboro, Oregon 97124-7055
USA

Photographs/Images — Credits/Copyrights

Page	Description	CREDIT/COPYRIGHT
2	Sasquatch model	Emmanuel Janssens
19	Albert Ostman	J. Green
20	Map	R. Patterson/author's collection
25	Certification: Naismith	J. Green
27	Ostman and Green	J. Green
28	Drawing: Ostman creature	I. Sanderson
30	Mount St. Helens	Google Earth © 2008, Digital Globe, TerraMetrics
32	Fred Beck	J. Green
33	Group: Mt. St. Helens	Public domain
33	Beck and Smith	Public domain
34	Ape Canyon	J. Green
35	Sketch: W. Roe	C. Murphy
36	Mount Mica	C. Murphy
36	Roe affidavit	J. Green
38	Map of California	D. Perez/author's collection
40	Jerry Crew	*Humboldt Times*
41	Book cover	R. Patterson/Hancock House Publishers
42	J. Crew and A. Genzoli	*Humboldt Times*
43	Tom Slick	L. Coleman
43	Peter Byrne	P. Byrne
44	Yeti cast	C. Murphy
47	Group: PNE	J. Green
48	Offield Mtn. full print	I. Marx
48	Offield Mtn. half print	I. Marx
52	Book cover	Pocket Books/P. Byrne
54	Bobcat drawing	Dick Barbre
56	*The Moment*	M. Rugg
56	Film site model	C. Murphy
57	Map	Y. Leclerc/C. Murphy
57	R. Patterson	R. Dahinden/Y. Leclerc
57	Bob Gimlin	C. Murphy

58	Frame 352: P/G film	R. Patterson (public domain)
62	R. Dahinden	C. Murphy
65	Boat with insignia inset	C. Murphy
66	Don Abbott	J. Green
67	Note (hotel clerk)	R. Dahinden
68	R. Dahinden with casts	J. Green
70	R. Dahinden and P. Byrne	C. Murphy
72	R. Dahinden, R. Patterson and van	J. Green
74	Map: Washington and area	Google Earth © 2008, TerraMetrics, Tele Atlas, Europa Technologies.
84	B. Heuvelmans	J-P. Debenat
86	Ed Fusch business card	E. Fusch
100	Grand Coulee Dam (looking east)	Public domain
100	Grand CouleeDam (aerial view)	Public domain
100	Grand Coulee Dam worker	Public domain
103	Farm-in-a-Day house	Public domain
111	Map: Washington	Author's collection
117	Bigfoot Exhibition trailer	P. Byrne
119	Postage stamp: Lewis and Clark	U.S. Postal Service
120	Salmon (art)	Author's collection
121	Native fishing	Author's collection
125	Salmon jumping	Author's collection
135	Celilo Falls and fishermen	Wikipedia (public domain)
135	Columbia River gorge	Wikipedia (public domain)
137	Sternwheeler *Okanogan*	Public domain
141	Postage stamp: Johnny Appleseed	U.S. Postal Service
149	Andy Joseph, Sr.	A. Joseph/author's collection
151	Colville Confederated Tribes insignia	D. Healy
161	Dmitri Bayanov and Jane Goodall	D. Bayanov
162	Grover Krantz	J. Green
164	Book cover: *Big Footprints*	Johnson Printing Co/GKrantz
164	Book cover: *Bigfoot/Sas. Evidence*	Hancock House Publishers (photo by John Cardinal)
168	Jambo statue	J-P. Debenat
171	Three skulls	G. Krantz
177	*Gigantopithecus* jaws	G. Krantz
177	Jaws	G. Krantz
178	Skulls	C. Murphy

178	*Gigantopithecus* drawing	Author's collection
179	*Gigantopithecus* head/shoulders	G. Krantz/John Cardinal
183	Hominoidea chart	Pascal Picq/author's collectio
184	Human phylogeny chart	Yves Coppens (author's collection)
191	M.J. Koffmann and J. Roumeguére	Author's collection
197	Professor Boris Porshnev	I. Bourtsev/D. Bayanov
198	Sleeping Ksy-Gyik	Author's collection
203	Zana: head and shoulders	B. Bannon
205	Khwit	I. Bourtsev/D. Bayanov
206	Igor Bourtsev and Khwit's skull	I. Bourtsev
206	Skull: Khwit (inset)	D. Perez
211	Human eye diagram	Author's collection
211	Human eye illustration	M.J. Koffmann
214	René Dahinden in Moscow	I. Bourtsev/D. Bayanov
215	Igor Bourtsev and Dmitri Bayanov	D. Perez
220	Aardvark	Wikipedia (see note p. 408)
226	Piotr Klafkowski	Author's collection
228	Marjorie Latimer and coelacanth	Author's collection
235	Giant panda	Wikipedia (see note p. 408)
235	Red panda	Wikipedia (see note p. 408)
237	Komodo dragon	Wikipedia (see note p. 408)
239	Orang-pendek drawing	A. Lindbergh
242	Book cover	Presses de la Cité/A. Lindberg
244	Face of *Homo Pongoides*	B. Heuvelmans
252	Evolution diagram	Yvette Deloison (author's collection)
255	Committee group	I. Bourtsev/D. Bayanov
256	Chris Murphy and skeleton	C. Murphy
256	Chris Murphy and skull	C. Murphy
257	Phylogeny	François de Sarre
258	Peter Byrne and J-P. Debenat	J-P. Debenat
259	Frame 352: P/G film	R. Patterson (public domain)
260	Study: Creature, P/G film	C. Murphy
261	Footprint casts: film site	C. Murphy
261	Footprint casts: cripple foot	C. Murphy
262	Bob Titmus with casts	J. Green
263	Hand print casts	C. Murphy
263	Knuckle cast	C. Murphy

264	Skookum cast and J. Meldrum	R. Noll
265	*Gigantopithecus* model with W. Munns	W. Munns
266	Grand Coulee prairie	J-P. Debenat
267	Durrell statue	J-P. Debenat
268	Zana and baby illustration	B. Bannon
269	Igor Bourtsev with skull	D. Perez
270	Julien Debenat and pandas	Julien Debenat
271	Dzonokwa carving (Vancouver)	J-P Debenat
272	Dzonokwa carving (Holm)	J-P Debenat
273	Native mask drawing	P. Travers
274	Native mask	C. Murphy
275	Gagiid carving (Bennett)	J-P. Debenat
276	Raven carving (Bennett)	J-P. Debenat
276	Whale carving (Bennett)	J-P. Debenat
277	R. Bennett	J-P Debenat
277	Hands working (R. Bennett)	J-P Debenat
278	Transformation mask (top image)	D. Hancock
278	Transformation mask (lower image)	D. Hancock
279	Native dancer	D. Hancock
280	Fred Bradshaw	J-P. Debenat
280	Nepalese mask	J-P. Debenat
281	Wild Men and Moors	Museum of Fine Arts, Boston, MA., USA.
282	Raven carving (W. Reid)	J-P. Debenat
283	Temple painting	Julien Debenat
284	Yeren painting	Author's collection
285	Ed Fusch	J-P. Debenat
285	Ed Fusch and author	J-P. Debenat
286	Ray Crowe	J-P. Debenat
286	Author with F. Bradshaw and W. Moore	J-P. Debenat
286	Author with Mary Dodds Schlick	J-P. Debenat
287	Author and Paul LeBlond	J-P Debenat
287	Author with professor, China	J-P Debenat
287	Author with interpreter, China	J-P Debenat
288	Portrait of B. Heuvelmans	J-P. Debenat
292	Nebuchadnezzar	W. Blake
293	Merlin the Enchanter vase drawing	Author's collection
296	Basque shepherds/"Basa Jaun"	Christian Le Noël
297	Postage stamp, bear	La Poste, France

298	Forest men of Transylvania	Author's collection
299	Bigfoot portrait	Stefano Maugeri
301	Wild man sculpture	Christian Le Noël
303	Book cover	University of Chicago Press/G.Schaller
316	Native basket	Public domain
323	Chief Joseph	Author's collection
334	Longhouse drawing	Author's collection
339	Graduation ceremony	J-P. Debenat
341	Sweat lodge	Author's collection
342	Book cover	Caxton Printers Ltd.
346	Killer whale artwork	Author's collection
351	Kwakiutl heraldic pole	J. Green
354	Bigfoot head	W. Moore
360	Mask, Kwakiutl	J. Green
361	Body painting illustration	Franz Boas
363	Kwakiutl mask sketch	R. Butler
378	Drawing by Fred Bradshaw	F. Bradshaw
379	Almasty drawing	M-J. Koffmann
379	Neanderthal depiction	Author's collection
380	Barmanu drawing	J. Magraner
380	Wild men of China drawing	Author's collection
381	Wild man church sculpture	Author's collection
382	Showman and his "beast" sculpture	Author's collection
383	Wild man with staff/shield sculpture	Author's collection
384	Wild woman and offspring sculpture	Author's collection
385	Romanic sculpture – bear and knight	Author's collection
386	"Ball of Burning Men" etching	Author's collection
387	"Game of Bear" actor	Author's collection
388	Cast and human foot comparison	J. Green
389	Bigfoot model by Emmanuel Janssens	E. Janssens
390	Sculpture, Colville Reservation	E. Fusch
390	Ed Fusch and author with sculpture	J-P. Debenat
391	Group photo, Mount Hood, 1994	J-P. Debenat
410	Debenat with bottle	B. West

NOTE (Pages 220, 235, 237): These images are from the Wikipedia Encyclopedia and are used with permission under the Creative Commons License. Please refer to the Wikipedia website for details.

APPENDICES

Jean-Paul Debenat, an assistant professor of English visiting from France to research Northwest animals in mythology, including the Sasquatch, was presented with a bottle of gamay noir, 1993, with the label, "Bigfoot Blend," by Ray Crowe, director of the Western Bigfoot Society in St. John's, OR. The caption on the bottle under the creatures reads: "They make this stomping grapes?" "And with those little feet!" the other exclaims. Photo by Bonnie West

Appendix 1

Article: *The North Coast News* (Grays Harbor, Washington), July 26, 1995.

Harbor visited during research for book

For most of the residents of Grays Harbor, the subject of Sasquatch comes up only occasionally in conversation and no amount of car mileage is spent traveling around the country trying to find him.

But a Frenchman with an interest in the creature has visited the Pacific Northwest not once, but three times, to gather research for a book he plans to write on the subject of Sasquatch, including it in a book of the Pacific Northwest's bestiary.

Jean-Paul Debenat, a good humor and friendly man who has been an assistant professor of English and Communications at the Institute of Technology of Nantes, France, for twenty years, had his interest in the creature piqued a number of years ago in England.

Debenat was invited to be one of the speakers at a three-day symposium held at the University of Surrey in England by the International Society of Cryptozoology, a group of researchers that deal with unknown and supposedly extinct animals.

"Cryptic means hidden, secret," he said, "and is the study of animals we don't see very often."

He said the term cryptozoology was coined in 1950 by Bernard Heuvelmans, a Frenchman who published a book on the subject, and one Debenat read that began his interest in the subject.

"Every year a new species is identified and acknowledged by the cryptozoologists," Debenat said, "but size-wise they don't amount to much. Insects, shells and birds make the headlines of scientific journals, not newspapers."

At the symposium, Debenat spoke on "Fabulous Beasts of Our Times," incorporating into his theme the great animal figures of European mythology and literature.

Through the centuries, Debenat said, any creature considered to be just too incredible to be believed was considered a fable, and anyone who insisted they existed was considered crazy.

As an example, he mentioned the Komodo Island dragon which still hadn't been identified in the early part of the twentieth century.

"A pilot crashed on the island and when he was rescued and taken back to the mainland he told people: 'I've seen a live dragon!' They locked him up believing he was suffering from the shock of his crash landing," he said.

Debenat was fascinated when he heard another speaker at the annual conference, Ed Fuchs of Riverside, WA, give a talk on the subject of Sasquatch, presenting it from a scientific point of view.

Back in France he discovered there wasn't material about the creature to study.

"The French people are just not very informed about him. There is only one book in France on the subject of Sasquatch and in that, only one chapter makes reference to him!" he said, raising his eyebrows as if to say, "Can you imagine that?"

Debenat first came to the States three years ago to accompany a group of computer science students from the institute who spent their ten day visit at the computer facilities at Microsoft, Boeing and Safeco.

He contacted Ed Fuchs, making a trip to visit the man at his Riverside, WA, home, to discuss Sasquatch.

In February 1994, Debenat, who also teaches English at the University of Nantes, was granted a six-month sabbatical to stay in Seattle as a visiting scholar.

He traveled to Seattle with Lauric Guillaud, a teacher of American literature at the university, and both undertook separate projects.

Debenat's was to study the main figures of the Pacific Northwest bestiary—orca whale, bear and eagles among them, that teach lessons in morality or philosophy.

In medieval times, he explained, it was a collection of tales in which the heroes were animals, and these animals were teachers.

They are like the animals in the native American culture, he explained, where the animal is a spiritual guide on a down-to-earth level.

Ed Fuchs was contacted again on this second visit, and helped

Debenat with his quest—helping him establish contacts with writers and critics in the Seattle area for his journal.

"When I began it was like some kind of initiation—a self-selected initiation and you don't realize it is going to be hard," he said.

Debenat was able to collect data for his research at the University of Washington where he found over 100 entries on the computer.

"There was a lot to study at the university," he said. "There were books, articles, and at least a minimum of two categories about Sasquatch.

"There was the physical side—sightings and evidence, and that which was intangible that had to do with ancient European mythology and old American Indian mythology."

One thing that stood out to Debenat was that the information collected over the years on the American Indians with regards to Sasquatch was not sufficient; but he did find that with the different languages among the tribes, it was more or less the same creature.

In the last three months of his research sabbatical in 1994, Debenat stayed with Ed Fuchs in Riverside.

There he was introduced to members of the Okanogan Nation and others who told him stories and sightings about Sasquatch.

This year, in the short two-week trip to the States Debenat took this month, he was able to talk about Sasquatch with Indians who live in Seattle, and traveled to Mt. Hood in Oregon to spend time at the Bigfoot Research Center headed by Peter Byrne.

Byrne gave Debenat access to his extensive computer files on Sasquatch.

He also spent time in the field of Grays Harbor with Elma resident, Fred Bradshaw, who took him to many Sasquatch sighting areas.

How did the instructor feel about getting instructions?

"I accept lessons from authorized people like Fred who has taken me in the field to help me understand more about Sasquatch," he said.

"And I accept lessons from authors Marjorie Halpin and Michael Ames who are the editors of *Manlike Monsters on Trial.*

"Halpin is the Curator of Ethnology at the Museum of Anthropology in Vancouver, BC, and Ames is the director of the museum. I respect their work on Sasquatch, it's serious stuff."

He said at this time in history there is a serious scientific interest in the creature, an underlying spiritual interest in him from white Americans who talk the most about Sasquatch, and an understanding among the Native Americans that Sasquatch will always be a part of their ancestral culture.

When asked if he would be returning to his Province of Brittany convinced of its existence he answered:

"I don't believe it is 100 percent real—but..." he said, making a "Who knows?" gesture with his hands and uplifted shoulders.

He sat quietly for a moment and added, "It's mysterious to me, but I tend to think it exists."

He said some of his colleagues in Nantes, which is Seattle's sister city, take him seriously and others, he said, say, "Well, he's a funny eccentric guy."

"But it's fun to take intellectual risks," he said. "Anyway," he added, chuckling that it didn't matter what they said, "I've got tenure."

France will soon become more familiar with the Pacific Northwest's Sasquatch. Debenat plans to give lectures on the subject there.

From his accumulated research, he is hoping to make a book on the bestiary of the Pacific Northwest available in France in about a year and a half.

And will he return to the Pacific Northwest again?

"I would love to. This study gave me the opportunity to discover the landscapes of the Pacific Northwest, the legendary Columbia River, and Eastern Washington," he said.

"And the people have always been very helpful and considerate toward my research."

Appendix 2

The Western Bigfoot Society

Among the various associations devoted to bigfoot investigations, Ray Crowe's Western Bigfoot Society (WBS) (created about 15 years ago and terminated in 2007), deserves special mention. Ray (now retired) managed a secondhand bookstore in the suburbs of Portland, Oregon. The WBS published a newsletter, *The Track Record* which appeared 10 months out of 12. On the front page of the *Record* was printed a warning: "WEAR SKEPTICALS WHILE READING PLEASE."

Ray published everything relating to the wild man, in the widest sense. Most of the articles reported bigfoot observations. However, there were items about African or Malaysian hominoids, about prehistoric human fossils, or about the discovery of new animals. As an example, an article shows black-and-white illustrations of skunk cabbage *(Lysichiton americanus)* drying on fallen tree trunks. Six parallel tree trunks are being used as a drying rack for the large leaves of this wild cabbage. The explanatory text quotes Nancy Turner's book, *Food Plants of British Columbia Indians* (1975), pointing out that this plant was only rarely used as food by the Indians. The Washington State Quinault tribe sometimes roasted the leaves to eat them. Unfortunately, sometimes children were poisoned; the plant like others of the arum family contains crystal of calcium oxalate which provokes violent irritations. Its roots should never be eaten raw. Once cooked, the leaves are edible, with a strong ginger-like taste. Bigfoot seems to like skunk cabbage.

Elsewhere in the *Record* appears a summary of an article dating from 1786, originally published in the *New Jersey Magazine* and *Monthly Advertiser*, describing the discovery of a wild man in the Pyrenees. The wild man had been discovered by shepherds. He was very tall and nimble, as hairy as a bear, but friendly and inoffensive. He would enter peasants' huts without stealing anything. He liked to run after the sheep for fun and scatter the flock. He appeared to be about 30 years old. The local forest spreads all the way into Spanish territory. Did the lone wild man come from Spain? Was he perhaps an abandoned child who might have survived by eating plants and roots? That was the question asked in the article by a Mr. Leroy, whose job was to procure tree trunks for the French navy.

In the October 1955 issue (No. 52) there is a letter from a Japanese correspondent, Yasushi Kojo, who reports on his analysis of three hair samples, one of which was provided by Fred Bradshaw. Regarding the black hair sent by Fred, the author concluded: "It is not the hair of an animal. It might come from a plant or be a synthetic fiber."

The WBS also holds monthly meetings which often attract gifted speakers—field workers, veterinarians, naturalists, geologists, writers, etc. Yearly, over the past dozen years, the WBS has organized an international congress in Hillsboro, near Portland. Speakers come from as far as France, Australia or South Africa. *The Track Record*, reported on the presentations in regular as well as special issues.

For example, *The Track Record* of May 2001 (No. 107) relates how an employee of the California Forest Service observed a bigfoot for two or three minutes near Almanor Lake. When he returned to the site on the following day with a pair of friends, he found two large prints as well as a 25-inch (65-cm) spear, or perhaps rather a club, as it was thick and heavy.

Ray Crowe welcomed contributions from a wide audience. It was up to the reader to select what was of interest. One found short pieces of news like the sighting of an abominable snowman by an officer of the military region of Chengdu, Sichuan, China. The officer sojourned in the Prefecture of Ngari, Tibet in 1960. A snowman tried to pull a rifle from the hands of a soldier who killed the creature. With the help of his comrades, he buried it without keeping any part of the unusual creature. Such reports, brief as they may be, demonstrate the degree to which the wild man is of interest everywhere.

In 2000, an imprint of a reclining bigfoot showing its hip, thigh and heel as well as the forearm on which it was leaning, was discovered in Skamania County, Washington near Skookum Meadows. Skookum also means "big" in the Chinook jargon, and hence also bigfoot: a powerful and evil forest god.

Dr. Fahrenbach, a researcher in the Oregon Regional Primate Research Center (now retired), examined 56 hairs gathered from the Skookum print. Casting aside rootlets, after identifying hair from deer, moose, bear and coyote, he focused on the scrutiny of a primate hair. Comparing it with six other hairs supposedly coming from bigfoot, he concluded that it belonged to the same category—that of "possible" bigfoot hairs.

A plaster cast was obtained by researchers, including Dr. Leroy Fish, a zoologist, who examined it using sophisticated laser mapping software. Anthropologist Jeff Meldrum, of Idaho State University estimates that the imprint is that of a hairy hominid more than eight feet (2.50 m) tall. The Skookum cast was the topic of many articles in *The Track Record* as well as in mainstream scientific magazines, for example the *New Scientist* (Vol. 168, No. 2270, 23 and 30 Dec 2000).[1]

Investigators searching for material evidence attach almost as much importance to the Skookum cast as they do to the Patterson/Gimlin film. The wealth of detailed information gathered from it (creature weight, height, relative dimensions of various body parts etc.,) is a clear sign of the passionate dedication of bigfoot researchers.

Data are accumulating from everywhere. Over the past dozen years, paleontology, genetics and archaeology have played an increasingly important role. While I admired Ray Crowe's openness to a wide range of ideas, I could tell that he had a certain aversion towards academics—PhDs were viewed with some suspicion. Nevertheless, he eventually welcomed their contributions; they showed that they could help in explaining the mystery, or at least to place it within a broader perspective.

Ray Crow's *Track Record* is sorely missed.

---

1. For illustrations and discussion on the Skookum Cast, see Chapter 5 of *Sasquatch: Legend Meets Science* by Jeff Meldrum (Tom Doherty Assoc., New York, 2006) or pages 145–151 in *Meet the Sasquatch*, by Chris Murphy (Hancock House Publishers, Surrey, B.C., 2004).

Appendix 3

African Hominids

A review originally published in *Cryptozoology*, 12, 1993–96, pp. 82–87 of *Dossier X: Les Hominidés Non Identifiés des Forêts d'Afrique (The X File: The Unidentified Hominids of the African Forests)*, by Jacqueline Roumeguère-Eberhardt, Robert Laffont, Paris, 1990.

Each of Roumeguère-Eberhardt's books—not to mention her noteworthy films on the Masai rites made for French television—is part of a whole. If the reader is kind enough to share this assumption with me, I will be justified in providing a brief survey of the author's earlier works before tackling the heart of the matter.

Let us begin with *Pensée et Sociétés Africaines: Essais sur une Dialectique de Complémentarité Antagonistique chez les Bantu du Sud-Est* (*African Thought and Society: Essays on a Dialectic of Antagonistic Complementarity Among the Bantus of the Southeast*, Publisud, Paris, 1986). These essays deal with an initiatic knowledge which is progressively unveiled as it becomes operational within our inner consciousness. They reveal the various aspects of a fundamental dialectic which originates from the masculine-feminine duality in a way such as these terms (feminine-masculine) appear as antagonistic and complementary; thus, the whole social and religious structure of the Bantus is based on this duality.

In her first essay, part of which was published in English under the title: "The Mythical Python Among the Vendas and the Fulanis: A Comparative Note" (Archiv für Völkerkunde, Vol. 12, Vienna, 1958), the author describes the domba, a traditional dance performed by the Vendas of Northern Transvaal. The dancers reenact a myth which may be summarized as follows: the python took a second wife, who was unaware of the real nature of her husband. During the day, she worked in the fields, and was prevented from going back to the village by the first wife. However, she found a pretext to return to the village; there, in the men's square, she saw the python, who fled in a fury and disappeared in the depths of Lake Fundudzi. There came a drought, and famine spread. Obviously, the python was claiming his younger wife, and he wished her to join him in the lake. So the royal maidens set to making the ritual beer, mpambo. When it was ready, the women took it to the lake as the men danced along and sang. The younger wife, bearing the beer pot, entered the lake and disappeared in it. Then the rain fell.

The domba is a fertility rite which renews the kingdom, secures the king's throne, and warrants plenteous harvests and fecund wives. The domba—literally, "the serpent which uncoils"— is God.

Originally, humankind and the whole of creation sat in the womb of the python who vomited them. This explains why the initiates say that novices under training are "sitting in the python's womb." The novices sit in the yard, surrounded by a fence which appears as the skin of the python. At the end of the domba, they will emerge from it as new creatures, as thorough persons. Python skins are placed around the fence, and a clay serpent hangs from the tree under which the novices dance. The novices thus learn that the fetus vomited at birth originates from two serpents within the mother's womb: one of the serpents "belongs to the Gods," the other "belongs to men."

In another version (Fulani) the python forbade his twin brother to marry a maiden without breasts; i.e., who did not attend yet the school of puberty. In other words, a nubile girl. The brother disobeyed. The python said that the young wife should never look at him. She too disobeyed. Realizing that she was watching him while he was drinking, the python, in a flight of fury, destroyed her hut, and then fled to the river Niger. The cattle, other animals, and finally the whole of creation followed him. However, the python showed some compassion: he told his brother to cut a twig, and to hit the animals with it. All the animals that the brother hit were saved. All the others disappeared in the water. This is why every prehistoric animal disappeared from the earth.

The mythical python is obviously linked with the notions of creation and procreation. A catastrophe occurs when the young maiden breaks a taboo; indeed a nubile girl is not allowed to take part in the dance of the python, or domba.

Still, the ritual offerings (of beer by the Vendas, and of milk and butter by the Fulanis) are meant to overcome the rupture. The secrets involved in these offerings are taught in these initiatic schools, culminating in the domba precisely. These rituals restore a unity which had been broken. Through the domba a dual—or dialectic—move is involved; on the one hand, the mythical reality becomes part of the social order; on the other hand, the social order falls into harmony with the mythical order.

The "masculine-feminine" antagonism underlies the religious and social structure of the Bantus. The author herself became an initiate through venda and tsonga teachings. She insists on the dynamics of such opposing principles, which are oriented towards creation and re-creation. Incidentally, as a domba initiate, the author (let's call her JRE) was given access to elements of knowledge which remain foreign to the lay person. For instance, JRE once asked about a musical instrument of the Lemba people called a deza. The matter, shape, colors, and various parts of the deza constitute symbols which reflect the mythology and social structure of the Lemba people. A Lemba priest revealed to JRE the "laws of the deza," initiatic laws that one may "read" on the instrument, but he would not do it in the presence of his own brother, who, as a Christian, had received a "modern" education at school and was not a domba initiate. Playing the deza means to create, and the sound itself is the newborn child who is crying. The "laws of the deza" operate at three levels: individual, familial, and cosmic. JRE was offered a deza by the Lemba priest, who also authorized her to use at will the information he had provided her.

These brief samples from *Pensée et Société Africaines* are far from paying tribute to the whole book. However the reader may trust the appreciation issued by the journal *American Anthropologist,* which hailed JRE's collection of six essays as "an excellent contribution to the French school of Social Anthropology." It also considered JRE's work as following the "tradition of Mauss, Levi-Strauss and Griaule."

This laudatory assessment would also apply to JRE's previous book: *Le Signe du Début de Zimbabwe* (*Zimbabwe's Sign of the Origins,* Publisud, Paris, 1982). This detailed and precise study is the result of 22 years of research on Bantu society. Although it deals with the qualitative rather than the quantitative, it is illustrated with numerous diagrams and charts—at times quite complex—showing cosmic organization, social hierarchy, parental systems, and, most intriguing, their interrelationships.

In Part I, the role of the muputu (translated by JRE as "totem") goes beyond that of the totem as studied precisely by Levi-Strauss (see *Le Totémisme Aujourd'hui,* Presses Universitaires de France, Paris, 1980). It then appears that the "totem" plays its vital part in everyday social life. It is real, not only intellectual; and, according to JRE, is bound to re-emerge under various forms, as the evolution of African countries already shows.

Part II concerns various myths as represented by the "bird of Zimbabwe," phallic clay or canvas dolls, sacred rocks (Rock of Marriage, Rock of Incest, etc.,…and Bushman cave paintings. JRE warns the reader that these are only bits and pieces from a vast initiatic knowledge which is acquired after many long years under the supervision of African masters.

Part III is about "real knowledge," and supplies "operating models": it explains how the dynastic and parental models, the cosmic and mythical models, operate and influence the basic social life of the Bantus. The book closes on similarities found within the Masai society in Kenya. These two volumes published by Publisud should be considered sound scholarly studies. However, the layman may feel reluctant to study them when confronted with some of their complexities. In this case, he should acquire JRE's *Les Maasaï* (*The Masai,* Berger-Levrault, Paris, 1984), an album illustrated by the best color photographs I have ever seen on the subject. The text is clear and straightforward, and provides accurate data, as may be expected, from an author who spent so many years among the Masai.

Quand le Python se Déroule (*When the Python Uncoils,* Robert Laffont, Paris, 1988), an autobiography of JRE, met with great success with the general public. I find it disconcerting because of its odd chronology, stylistic weaknesses, and abundance of anecdotes. Perhaps this book should be approached as we would a sculpture of the Makonde people: typically, such sculptures display intertwined bodies and limbs carved in hard ebony wood. According to the Makonde artists, those weird shapes are combined "so that you may dis-understand." Nevertheless, on page 10, in a prose poem (Prelude), reproduced below, JRE mentions X, an unidentified hominid of the Equatorial forests, bound either to live a parallel life or to disappear (my own literal translation):

> Mr X, our brother, our father, our prototype,
> homo abilis or homo creatus,
> Or then homo unknown, but homo yet,
> Whatever the specialists may say or think.

Thus, in the opening pages of the book, the reader is confronted with the extraordinary Mr X "who lives next to us on parallel wavelengths." Then Mr X vanishes until page 229: there, a little more than one page is devoted to the hominid. Fortunately, JRE's latest work, *Dossier X,* is more apt at satisfying our curiosity, although only partly.

The book consists mainly of 31 accounts selected from a corpus provided by 190 observers (including two Europeans, a Swede and an Englishman). Each observer is given a code number (the Englishman is known as observer 35), just like the 22 forests in which Mr X has been encountered. JRE took this precaution so as to avoid the invasion of Kenya by journalists and tourists. It would be tiresome to sum up the various accounts, especially as they are presented with an apparent lack of logical order. Oddly enough, one is likely to derive some pleasure from those repetitious reports, which basically tell the same stories. Besides, Bernard Heuvelmans' lengthy foreword (29 pages) serves as a useful guide, considering the intricacies of the text. Heuvelmans (we shall "code" him BH) lists five types of hominids, summing up their main features:

X1: The tallest of them all, whose height ranges from 5 feet, 11 inches (1.8 m) to 8 feet, 2 inches (2.5 m). Known by the natives as "the one covered with moss," due to the short gray hair (reddish brown for the younger ones) on its body. Footprints are larger than those of a human. Hefty, peaceful, innocuous, and curious, X1 may seize a child or an adult, examine him/her and let him/her go unharmed. Armed with a club (a tree log or a branch), X1 will knock down a buffalo, drink the blood from the jugular vein, eat the bowels, and even break off one whole foreleg and carry it on its shoulders. Among various hypotheses, BH suggests that X1 could be a specific African *Australopithecus,* adding that the discovery of a living specimen would be of the highest value to understanding the evolution of primates and the problem of our own origins.

X2: Very tall, but not hairy. Lightly colored skin (the Masai call him *naibor,* "the White"). Somewhat aggressive; campfires trigger its anger, causing it to rush at them, and scatter the embers with a stick while yelling. Unlike X1, it does not live in the forest, but in caves. Aside from its height, X2 resembles a Bushman similar to those who lived in East Africa thousands of years ago.

X3: Looks like a negative proof of X2. Dark skin, long hair (down to the thighs), and white due to its age: almost 100 years. Hunts, kills, and eats buffaloes in the same manner as X1. Speaks the Masai language. BH thinks X3 could be a solitary runaway warrior, a Masai delinquent.

X4: No taller than 4 feet, 3 inches (1.3 m). Dark skin, big head, short fuzzy hair, strong chest and neck, short bulky arms, and long sharp nails. Sometimes seen wearing a cape made from some animal hide. Feeds on roots and tubers that it unearths using a plain wooden stick. Also eats mushrooms and is quite fond of honey. Does not hunt, but steals meat, eaten raw since it is ignorant of fire. In Tanzania, X4 is said to build small huts in the trees. Speaks a language connected with the Bantu linguistic group. Fears other men, and runs away from them. BH compares X4 to the homuncules described in his book *Les Bêtes Humaines d'Afrique,* although X4 would better be described as a "proto-pygmy."

X5: It is not so much an enigma, as , like X3, an individual case. X5 wears an animal hide, as well as laced sandals in the Roman fashion, carries a bag, a walking stick, a small bow, and poison arrows. JRE managed to purchase a bow, arrows, a quiver, and a bag from a native informer. Nobody either in Kenya or at the British Museum could identify X5's bow and arrows. Moreover, X5 uses a thick string made of elephant tendon, while the Ndorobos of the East African forests use zebra tendon. As a result, not a single Masai warrior was even able to bend the bow. However puzzling, X5, short, bald and elderly, is considered as yet another outcast by BH. Still, JRE believes that the capture of X5, who manages to survive incognito, would prove that unidentified hominids may easily hide in Kenya's forests.

BH, in spite of the affection and respect he bears to the author, remains critical towards JRE's text. He suggests, nonetheless, that the facts are presented in the African manner: pell-mell. "It is like an esoteric message from another world, very remote and at the same time secretly buried deep down in our inner selves." This book raises many questions concerning the real existence of Africa's hominids, the mythopoietic or myth-making process— here applied to the archetypal Wild Man—inherent in human nature, and the personality of JRE herself, whom BH, donning Trader Horn terminology, compares to the White Goddess.

In my own view, three tightly knit domains ought to be considered: the African background (which JRE described in her earlier works), the hominids (both under their "scientific" and archetypal aspects), and JRE with her particular approach. They should not be separated from one another. While observing how they influence each other, we might, in the long run, dis-understand, and at least learn something.

General Index

Abominable Snowmen of America
 Club, 53
Acheulean tools, 174
African Hominids, 418ff
almasty, 192, 212, 379
Ames, Michael, 215, 413
Ape Canyon, 34
Appleseed, Johnny, 140
Australopithecus 165, 171, 179, 183,
 238, 245, 249

Barmanu, 375, 380
Basa Jaun, 296
Bayanov, Dmitri, 203, 213, 214, 216,
 217
 photo, 161, 215, 255
bear rituals
 bear dance, Rumania, 297
 bear jig, Canada, 297
 model for shaman, 364
 mysterious herbalist, 311
 Prats de Mollo, 298, 387
Beck, Fred, 30
 photo, 32, 33
Bennett, Ralph. *See* Goolaslacoon
bigfoot
 and Coyote, 342
 at Wounded Knee, 325
 Chippewa legends, 353
 everywhere, 350ff
 guardian spirit, 311, 345
 model, 389
 name, 37
 portrait, 299
 fictional, 369, 373
Bigfoot Daze (1994), 367
Bigfoot Research Project, 52, 116, 413
bipedalism, 251
Bluff Creek, 38, 55
Boas, Franz, 360
bonobo, 165, 250
 Kanzi, 165

Bourtsev, Igor, 205, 214, 216
 photo 206, 215, 269
Bradshaw, Fred, 157, 367 , 378, 413
 photo 280
buck'was mask, 274
Buffetaut, Eric, 191, 373
Byrne, Peter, 43, 48, 69ff, 116, 413
 photo, 43, 70, 258

Capart, Andre, 223, 247
Cascade Mountains
 role in climate, 89
 geology, 90, 95, 115
Celilo Falls, 120, 132
 photo, 135
Chief Joseph, 95, 322, 331
 photo, 323
coelacanth, 61, 228ff
Columbia River, 81
 Bridge of the Gods, 119
 irrigation, 81
 dams, 98, 101, 334
 electrification, 106
Colville tribes, 95, 110, 317
 Museum, 148
Committee for the Study of the Question of the Snowman, 201, 254
 group photo, 254
Coon, Carlton, 246
Coppens, Yves, 169, 171, 184, 189
Courtenay-Latimer, Marjorie, 223,
 228ff
 photo, 228
Coyote, 120, 151, 304, 305, 307, 320,
 330, 332,333, 336, 339, 342, 362
Crew, Jerry, 39
 photo, 40, 42
Cro-Magnon Man, 175, 188
Crowe, Ray, 174, 415
 photo, 286
cryptozoology, 219
Cuvier, Baron, 233

Dahinden, René, 48, 62ff, 214, 223, 367
  photo, 62, 68, 70, 72, 214
Dalles (The), 115
dances, 303, 360
David, Father Armand, 233
dehumanization, 249
Debenat, Jean-Paul
  photo, 258, 287, 410
deBonis, Louis, 185
Deloison, Yvette, 251, 252, 255
De Sarre, François, 250, 255
DNA genealogy, 188
Dunn, Donald, 102
Dzonokwa, 271, 272, 278, 351, 357ff, 360

Ellul, Jacques, 188

Fabulous Beasts Conference, 84
fox, 337, 340, 354
Flores Island "hobbits", 239
Fredin, Adeline, 312, 314, 338ff
Fuchs, Ed, 85, 88, 106, 289, 314, 412
  photo, 285

Gagiid, 275, 365
George, Tillie, 312
giants, 350ff, 363ff
Giganthopithecus, 176, 178, 183, 191, 254
  model, 179, 265
Gill, George, W. 217
Gimlin, Bob, 53
  photo, 57
Goodall, Jane, 159, 160
  photo, 161
Goolaslacoon, 276, 277, 347
gorilla, 170, 178
  Jambo, 167
  Koko, 165
Gould, Stephen J., 253, 300, 302, 373
Grand Coulee, 92, 98ff, 101, 314
Green, John, 19, 25, 31, 38, 48, 65, 216, 350, 367
  photo, 27

Greenwell, Richard, 223, 224
grizzly, 308, 310
Guillaud, Lauric, 13, 191, 332, 412
Guthrie, Woody, 81, 102

Halpin, Marjorie, 215, 413
Hélias, J.-P., 25
Heuvelmans, B., 14, 86, 191, 194, 219ff, 223, 227, 231, 233, 237, 238, 241, 248, 252, 289, 290, 299, 369, 376, 411, 422
  photo 84, 288
Hilbert, Vi, 312, 335ff
Hillary, Edmund, 46
Holm, Bill, 363
Homo
  abilis, 183
  erectus, 171, 179
  ferus, 199
  floresiensis, 238, 239, 253
  pongoides, 241ff, 247, 376
  neanderthalensis, 172, 175, 180, 195, 206, 245, 247, 253, 291, 379
  sapiens, 171, 184, 206, 239, 249, 252
  troglodytus, 199

ice-man. See *Homo pongoides*
International Society of Cryptozoology, 10, 84, 223ff, 290
  directors, honorary members, 223

Janssens, Emmanuel, 389
Joseph, Andy, 148, 311, 314, 337
  photo, 149

Khakhlov, Vitali A., 196
Khwit (son of Zana), 204, 269
  photo 205
killer whale 276, 344ff
Klafkowski, Piotr, 225
  photo 226
Koffman, Marie-Jeanne, 191, 202, 209, 223, 254, 379
  photo 191, 255
Kokopelli, 336

426

Komodo dragon, 236
  photo, 237
Krantz, Grover, 71, 154, 162, 169, 170, 173, 176, 177, 185, 191, 193, 218, 254, 367
  photo 162
Ksy-Gyik, 198

LeBlond, Paul, 223
  photo, 287
Leenhardt, Maurice, 359
Lévi-Strauss, Claude, 357, 358, 366
Lewis & Clark, 83, 119, 306, 310, 314, 317ff, 332
Lindbergh, Alika, 221, 239, 241, 246
Linnaeus, 199
Lucy, 171, 187. *See also* Australopithecus

Mackal, Roy, 223, 290
Man-like Monsters on Trial, 215
Margraner, Jordi, 375, 380
Markotic, Vladimir, 298
Maugeri, Stefano, 299
Meldrum, Jeff, 264, 417
Menzel's experiment, 166
Merlin the Wild, 293, 294
Messner, Reinhold, 374
Mousterian culture, 205
Munns, William, 265
Murphy, Christopher, 56, 256
  photo 256
mythology, 289ff , 327

Napier, J., 29, 36, 47, 59,67, 223
Neanderthals. *See Homo neanderthalensis*
Nessie, 224
Nez Perce tribe, 310, 314, 317ff, 322
night vision, 209
Nokia, Bride of Bigfoot, 309

Okanogan County, 110, 137, 144
Okanogan Smith, 137
okapi, 231
orang-pendek, 238, 239

orangutan, 177, 179
orca woman tale, 345

Ostman, Albert, 19ff
  photo, 19, 27

Pan, 291, 298
panda
  greater, 234, 302
  lesser, 236
  photos, 235, 270
Patterson, Roger, 30, 37, 41, 53
  photo 41, 57, 72
Patterson–Gimlin film, 53ff
  Disney studios' view, 69
  frame, #352, 58, 259, 260
  Glickman's examination, 70
  in Moscow, 214
  Krantz's opinion, 71
  Napier's analysis, 59
Pichon, Jean-Charles, 13, 155, 159, 160, 330
Pickford, Martin 187
Picq, Pascal, 176, 182, 238
Pilou, 359
Pithecanthropus, 181, 237, 245
  swimming, 250
platypus 233
Porchnev, Boris F., 195, 196ff, 205, 247, 254
  photo 197, 255
potlatch, 357ff
Przewalski, Nikolai M., 233
Pyle, Robert, 77, 109

Qanelekak, 348

Raven, 276, 282, 354, 366
Reid, Bill, 282, 366
Rines, Robert, 224
Roberts, Richard, 238
Roe, William, 35
Roumeguère-Eberhardt, Jacqueline, 191, 418
  photo, 191

salmon, 120, 122
Ichtyoid fantasy, 126
Kwakiutl poem, 320
native rites, 132
speciation, 232
salmonberry tale, 357
Sanderson, Ivan T., 28, 37, 38, 243,
sasquatch
   ancestry, 173, 176, 257
   appearance, 26, 55, 59, 157
   benevolence, 27
   drawings, 28, 35, 56, 354, 378
   footprints, 48, 66, 67, 68, 113, 217, 261, 262, 388
   hand print, 263
   language, 22, 28, 166
   locomotion, 36, 55
   mask, 273, 274, 280
   phylogeny, 182
   salmon eaters, 334, 344
   using blankets, 22
   vegetarian, 22, 29
   vocalizations, 76, 77, 78, 216

satyrs, 291
Schaller, George, 302
Scott, Sir Peter, 223, 224
serpent worshippers, 329
shaman, 367
sightings
   Bluff Creek (1966), 53
   Bluff Creek (1958), 37
   Grays Harbor County, 156, 411
   Mount St. Helens, Beck, 30
   Mount St. Helens, Pyle, 77
   Neah Bay, 75
   Northern California (1972), 76
   Ostman's, 19
   Swenatum, 152
   William Roe's, 35
Sinanthropus, 237
Skookum Cast, 264, 417
skunk ape, 365
Slick, Tom, 43, 47
   photo, 43
   Pacific Northwest Expedition, 48, 65

Slick and Johnson expeditions, 43, 45
Smith, Hiram F. (Okanagan), 137
Smith, J.B.L., 61, 229
snowman, 196, 201
St. Césaire Man, 206
Streamer, Francis, 139, 143ff,
sweat lodge, 304, 305, 341
Swedenbord, E., 140
Swenatum, 150, 152
Sykes, Brian, 188

Titmus, Bob, 48
   photo, 262
Track Record (The), 415ff

Vercors, 12

Wallace, Ray, 39
West, Bonnie, 156
Western Bigfoot Society, 415ff
wild man, 12, 281, 283, 284, 291, 293, 301, 368, 380ff
   mask, 363
Windigo, 351
Wounded Knee, 325

X (Kenyan hominids), 192, 418ff

yeren, 284
yeti, 43, 183, 255, 374

Zana the Ogress, 203, 268

More HANCOCK HOUSE cryptozoology titles

Best of Sasquatch Bigfoot
John Green
0-88839-546-9
8½ x 11, sc, 144 pages

**Sasquatch:
The Apes Among Us**
John Green
0-88839-123-4
5½ x 8½, sc, 492 pages

In Search of Giants
Thomas Steenburg
0-88839-446-2
5½ x 8½, sc, 256 pages

Sasquatch Bigfoot
Thomas Steenburg
0-88839-312-1
5½ x 8½, sc, 128 pages

www.hancockhouse.com

More **HANCOCK HOUSE** cryptozoology titles

Bigfoot Encounters in New York & New England
Robert Bartholomew
Paul Bartholomew
978-0-88839-652-5
5½ x 8½, sc,
176 pages

Bigfoot Encounters in Ohio
C. Murphy, J. Cook, G. Clappison
0-88839-607-4
5½ x 8½, sc,
152 pages

Bigfoot Film Controversy
Roger Patterson, Christopher Murphy
0-88839-581-7
5½ x 8½, sc,
240 pages

Bigfoot Sasquatch Evidence
Dr. Grover S. Krantz
0-88839-447-0
5½ x 8½, sc,
348 pages

The Locals
Thom Powell
0-88839-552-3
5½ x 8½, sc,
272 pages

www.hancockhouse.com

Raincoast Sasquatch
J. Robert Alley
978-0-88839-508-5
5½ x 8½, sc,
360 pages

Hoopa Project
David Paulides
0-88839-653-2
5½ x 8½, sc,
336 pages

Tribal Bigfoot
David Paulides
978-0-88839-687-7
5½ x 8½, sc,
336 pages

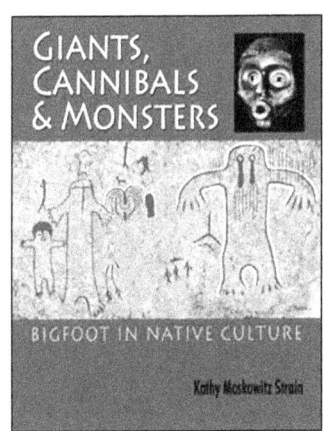

Giants, Cannibals & Monsters
Kathy Moskowitz Strain
0-88839-650-3
8½ x 11, sc, 288 pages

Who's Watching You?
Linda Coil Suchy
978-0-88839-664.8
5½ x 8½, sc, 408 pages

www.hancockhouse.com

More **HANCOCK HOUSE** cryptozoology titles

Know the Sasquatch
Christopher Murphy
978-0-88839-657-0
8½ x 11, sc, 320 pages

Meet the Sasquatch
Christopher Murphy,
John Green,
Thomas Steenburg
0-88839-580-9
8½ x 11, hc, 240 pages

Bigfoot Film Journal
Christopher Murphy
0-88839-658-7
8½ x 11, sc, 106 pages

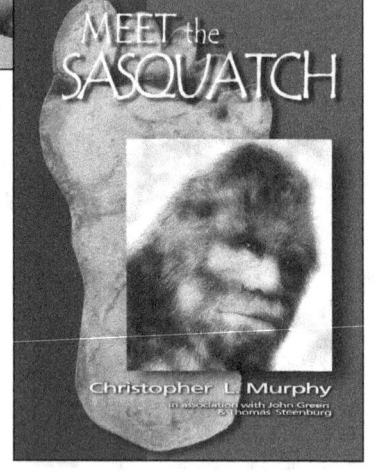

www.hancockhouse.com

www.ingramcontent.com/pod-product-compliance
Lightning Source LLC
Chambersburg PA
CBHW060936230426
43665CB00015B/1969